W9-BRJ-811

Beginning Hydroponics

Soilless Gardening:

A Beginner's Guide to Growing Vegetables, House Plants, Flowers, and Herbs without Soil

Updated with New Sources

Also by Richard E. Nicholls

The Running Press Book of Turtles

Beginning Hydroponics

Soilless Gardening:

A Beginner's Guide to Growing Vegetables, House Plants, Flowers, and Herbs without Soil

Updated with New Sources

by

Richard E. Nicholls

Running Press
Philadelphia, Pennsylvania

Copyright © 1977, 1990 by Running Press. All rights reserved under the
Pan-American and International Copyright Conventions. Printed in the
United States of America.

9 8 7 6
Digit on right indicates the number of this printing.

International representatives: Worldwide Media Services, Inc.,
30 Montgomery Street, Jersey City, New Jersey 07302,

Canadian representatives: General Publishing Co. Ltd.,
30 Lesmill Road, Don Mills, Ontario M3B 2T6.

Library of Congress Cataloging-in-Publication Number
89–43022

ISBN 0–89471–742–1 Library binding
ISBN 0–89471–741–3 Paperback

This book may not be reproduced in whole or in part in any form or by any
means, electronic or mechanical, including photocopying, recording, or by any
information storage and retrieval system now known or hereafter invented
without written permission from the publisher.

Cover art direction and design by James Wizard Wilson
Cover illustration by Charles Santore
Interior design by Peter J. Dorman
Interior illustration by Julian Kernes
Edited by Alida Becker

Typography: Times with Triumvirate by Richard Conklin,
Philadelphia, Pennsylvania.
Printed and bound by Port City Press, Baltimore, Maryland.

This book may be ordered directly from the publisher.
Please add $2.50 for postage and handling.

But try your bookstore first.

Running Press Book Publishers
125 South Twenty-second Street
Philadelphia, Pennsylvania 19103

Premise

Hydroponics, a method of gardening without soil, has developed over the past fifty years. While it has been proven both reliable and effective, and has received much attention in some quarters, it is still an unfamiliar practice to many gardeners.

This book is designed to introduce you to the principles and methods of hydroponic gardening.

Contents

Introduction

Hydroponics is the term used to describe the several ways in which plants can be raised without soil. These methods, also known generally as soil-less gardening, include raising plants in containers filled with water or any one of a number of other nonsoil mediums—including gravel, sand, vermic-ulite—and other more exotic mediums, such as crushed rocks or bricks, shards of cinder blocks, and even styrofoam.

There are several excellent reasons for replacing soil with a sterile medium. Soil-borne pests and diseases are immediately eliminated, as are weeds. And the labor involved in tending your plants is markedly reduced.

More important, raising plants in a nonsoil medium will allow you to grow more plants in a limited amount of space. Food crops will mature more rap-idly and produce greater yields. Water and fertilizer are conserved, since they can be reused. In addition, hydroponics allows you to exert greater con-trol over your plants, to insure more uniform results.

All of this is made possible by the relationship of a plant with its growing medium. It isn't soil that plants need—it's the reserves of nutrients and moisture contained in the soil, as well as the support the soil renders the plant. Any growing medium will give adequate support. And by raising plants in a sterile growing medium in which there are *no* reserves of nutrients, you can be sure that every plant gets the precise amount of water and nutrients it needs. Soil often tends to leach water and nutrients away from plants, making the application of correct amounts of a fertilizer very difficult. In hydroponics, the necessary nutrients are dissolved in water, and this resulting solution is applied to the plants in exact doses at prescribed intervals.

This book is intended as an introduction to the practice of hydroponic gardening. After discussing the history and development of this alternative method of gardening, and some of the reasons why hydroponics may become increasingly important to us in the near future, the various methods of hydroponic gardening are discussed. The specifics of operating a unit are covered, and several plans for simple hydroponic units are given. Almost any kind of plant that can be grown in soil can be grown in a hydroponic unit, but those kinds of plants best suited to hydroponic culture are mentioned. Pest, disease, and nutritional problems that may occur are also discussed, and remedies suggested.

A Resources section lists suppliers in the field; and the Bibliography lists other books for those interested in pursuing the subject.

* * *

The essential principles of hydroponics have been known for over one hundred years, but it was only very recently that anyone realized their tremendous importance as a method of gardening. Until 1936, raising plants in a water and nutrient solution was a practice restricted to laboratories, where it was used to facilitate the study of plant growth and root development. In that year, Dr. W.F. Gericke, working in the laboratories of the University of California, succeeded in growing twenty-five-foot tomato vines in large basins containing a water and nutrient solution. It was Dr. Gericke who invented the name for his discovery. Formed from two Greek words, hydroponics can be roughly translated as meaning "water working."

Gericke's experiments caused a sensation. Photographs of the professor standing on a stepladder to gather in his crop appeared in newspapers throughout the country, along with articles hailing the discovery in the most outlandish manner. Hydroponics was said to be among the most important discoveries of the twentieth century and would make traditional methods of farming obsolete. During the brief but highly enthusiastic period immediately following the announcement of Dr. Gericke's achievement, promoters exploited the public's interest by peddling useless materials passed off as complete hydroponic growing units. Disillusionment and the Depression soon pushed hydroponics out of the spotlight. But if the professor's accomplishment did not have any immediate lasting effect on the public, it did set off a number of projects in laboratories.

Between the time of his discovery and the onset of American involvement in World War II, Dr. Gericke and a number of other scientists were able to develop hydroponics from a concept into a sophisticated, successful system. Indeed, its principles became so well understood that American soldiers stationed on a number of harsh, isolated islands in the Pacific were able to raise all of their own vegetables in greenhouses rigged to grow crops hydroponically. Large-scale hydroponic greenhouses were set up in Japan directly after the war to feed American occupation forces.

Since then, many hydroponic units, some on a very large scale, have been built in Mexico, Puerto Rico, Hawaii, Israel, Japan, India, and Europe. In the United States, without much public awareness, hydroponics has become big business. Thousands of small hydroponic units and many hydroponic greenhouses are now in operation. The government funds a variety of experiments in the field, private institutions carry out research projects, and numerous firms are involved in manufacturing hydroponic units or supplies for such units.

Hydroponics is an idea whose time has finally come. As interest in it revives, there is now sufficient knowledge and a wide enough selection of equipment to sustain public interest and prevent disillusionment. More than fifty years after its discovery, hydroponics has become a sophisticated, eminently successful science, a viable alternative method of growing plants. It has also, thanks to a tremendous amount of research, become simplified to the extent that even someone with no previous gardening experience can easily raise plants in a hydroponic unit.

But why should we need it? What makes it more than a fascinating novelty?

If soil was everywhere of consistently high quality, and if all who wanted to grow plants had the ground in which to do so, perhaps hydroponics would seem no more than a curious experiment. But in fact soil varies greatly in its consistency and quality. When we plant outside, even the best soil must be thoroughly treated with fertilizer. Many of us, as dwellers in cities or in suburban apartments, lack even the smallest plot of ground on which to raise plants. We can of course raise some plants in pots or tubs filled with soil, but this has always been a slightly messy business.

The principles behind hydroponics are eminently adaptable. Even if you live in the tiniest of apartments, you can raise flowering plants, fruits, herbs, and vegetables in a hydroponic unit. You can raise plants hydroponically in individual pots or containers, or in units tailored to fit whatever size and shape of space you have available. If you live in a city, or in any situation in which you do not have access to a plot of ground, a hydroponic garden is the best—and indeed the only logical—alternative.

Even if you have enough land, there are compelling reasons to consider growing at least some of your plants, fruits or vegetables hydroponically. A hydroponic unit requires only about twenty minutes of care a day. Even a greenhouse converted to hydroponic cultivation requires much less care than a greenhouse outfitted with soil-filled trenches. Fruits and vegetables grown in a hydroponic unit are generally superior in flavor, appearance, and nutritive value. In addition, you can produce crop after crop with consistently excellent results.

The primary concern of every kind of plant cultivation has always been control. The development of fertilizers, insecticides, and hardier strains of plants has all been due to the desire to exert a greater control over the outcome. Until recently, the greenhouse was as far in this direction as we could go. It formed a compact environment in which almost every element affecting plant growth could be regulated. But even in a greenhouse certain problems remain, soil-borne diseases and pests being among the most persistent.

Outdoors, there are the additional problems of changing weather conditions, and the difficulties inherent in a garden—the constant necessity of weeding and working the soil to replenish the nutrients needed by the plants. Adding nutrients to soil is always a rather inexact effort, for some of what you add is bound to be leached off. Because plants grown in a hydroponic unit can be given very exact doses of nutrients and will absorb all of what they are given, the crops you raise hydroponically will develop to optimum levels of appearance and taste. Because the roots of plants grown in a soilless medium don't need to constantly grow in a search for nutrients, more plants can be grown in a smaller area. You will thus be making the best use of whatever space is available to you. You'll also be able to grow plants faster—again, because you have removed so many of the uncertainties usually attendant upon gardening. Growing plants in a soilless medium eliminates the possibility of soil-borne pests or diseases ravaging your crops or plants. It also eliminates the need for you to rely on powerful insecticides.

Despite these advantages, hydroponics will never replace conventional methods of agriculture. Many crops, such as the cereals or corn, are not easily adaptable to culture in hydroponic units. On a more aesthetic level, it should be noted that many people simply enjoy the tasks associated with gardening outdoors.

However, hydroponic gardening is mightily attractive. It allows you to grow plants or crops of optimum quality at a low cost with less effort than is required for raising plants in soil, to grow them faster, and to grow them largely free of the common problems of pests and diseases. Hydroponics is also very versatile. You can apply its principles to raising plants in individual pots, in small groups, or in special, easily constructed units. You can also turn a greenhouse into one large hydroponic unit and make most of the care of the unit automatic, regulated by gauges and machines. Whatever your needs, whatever your budget, hydroponics can be adapted to meet them.

There is a procedure in every part of nature that is perfectly reg-ular and geometric, if we can but find it out; and the further our searches carry us, the more we shall have occasion to admire this and the better it will compensate our labours.

John Woodward,
*Some Thoughts and Experiments
Concerning Vegetables,* 1699

1. The Science of Hydroponics

THE HISTORY OF HYDROPONICS

Some three centuries ago, John Woodward, an English scientist and a fellow of the Royal Society, undertook the first recorded scientific experiments on the subject of plant nutrition. He wanted to know whether plants drew nourishment from the soil or from water. Woodward discovered that adding small amounts of soil to the water in which plants were being grown caused the health of the plants to improve. As a result, he theorized that it was soil, and not water, from which a plant drew nourishment.

His findings flatly contradicted a belief, which had been dominant throughout the Middle Ages and up to his time, that in plant growth "the water is almost all in all. . . . The earth only serves to keep the plant upright." This belief was obviously based on the centuries of experience Western farmers had had with drought. When they could no longer deliver water to their plants, the plants died, no matter how rich the soil was.

Woodward's work set off a good many more experiments on the still perplexing matter of plant nutrition. While the question of the source of plant nutrients was not finally resolved until the middle of the nineteenth century, it was preceded by several equally important related discoveries. These discoveries, together with the determination of the source of plant nutrients, made possible the development of the science of hydroponics.

In 1792 the brilliant English scientist Joseph Priestley discovered that plants placed in a chamber having a high level of "fixed air" (carbon dioxide) will gradually absorb the carbon dioxide and give off oxygen. Jean Ingen-Housz, some two years later, carried Priestley's work one step further, demonstrating that plants set in a chamber filled with carbon dioxide could replace the gas with oxygen within several hours if the chamber was placed in sunlight. Because sunlight alone had no effect on a container of carbon dioxide, it was certain that the plant was responsible for this remarkable transformation. Ingen-Housz went on to establish that this process worked more quickly in conditions of bright light, and that only the green pads of a plant were involved.

In 1860, Professor Julius von Sachs published the first standard formula for a nutrient solution that could be dissolved in water and in which plants could be successfully grown. This marked the end of the long search for the source of the nutrients vital to a plant. It had been established that a plant absorbed water through its roots and then sent it upwards to be distributed throughout the plant. The nutrients necessary to plant growth were found to be held in the soil. The application of water to soil caused these nutrients to be dissolved and thus carried into a plant suspended in the water. Plants were found to possess the capacity to extract these nutrients from the water they absorbed.

Curiously enough, no one at that time realized the potential applications of these discoveries. Growing plants in a water and nutrient solution was seen merely as a technique for use in a laboratory situation to facilitate the study of plants and nutrients. Not until 1936 were the implications for agriculture made plain. In that year, W.F. Gericke and J.R. Travernetti of the University of California published an account of the successful cultivation of tomatoes in a water and nutrient solution.

In Gericke's method, plants were placed in waterproof basins on screens of wire mesh. The roots grew through the mesh into a water and nutrient solution that filled the bottom of the container. Between the mesh and the liquid, some space was left so that the roots could also receive oxygen. This method required a degree of skill and experience that few possessed, so the initial excitement over hydroponics rapidly subsided. However, it had served some purpose. A number of commercial growers started experimenting with the technique, and researchers and agronomists at a number of agricultural colleges began working to simplify and perfect the procedures.

Their efforts paid off in everything but publicity. Partially because of public disenchantment with hydroponics and also because of the grim nature of the times, all mention of hydroponics disappeared from the newspapers that had once touted it as the salvation of a hungry world. But the work went on.

Ironically, neither Gericke nor any of the determined scientists preceding him can be said to be the discoverers of the principles of hydroponics. They formulated them, they proved them, they applied them, but even centuries before John Woodward's experiments, the same principles were being successfully applied.

The best example that comes to mind is that of the Aztecs of Central America. A nomadic tribe, they were driven onto the marshy shore of Lake Tenochtitlan, located in the great central valley of what is now Mexico. Roughly treated by their more powerful neighbors, denied any arable land, the Aztecs survived by exercising remarkable powers of invention. Since they had no land on which to grow crops, they determined to manufacture it from the materials at hand.

In what must have been a long process of trial and error, they learned how to build rafts of rushes and reeds, lashing the stalks together with tough roots. Then they dredged up soil from the shallow bottom of the lake, piling it on the rafts. Because the soil came from the lake bottom, it was rich in a variety of organic debris, decomposing material that released large amounts of nutrients. These rafts, called *chinampas,* had vegetables, flowers, and even trees planted on them. The roots of these plants, pushing down towards a source of water, would grow through the floor of the raft and down into the water.

These rafts, which never sank, were sometimes joined together to form floating islands as much as two hundred feet long. Some *chinampas* even had a hut for a resident gardener. On market days, the gardener might pole the raft close to a market place, picking and handing over vegetables or flowers as shoppers purchased them.

By force of arms, the Aztecs defeated and conquered the peoples who had once oppressed them. Despite the great size their empire finally assumed, they never abandoned the site on the lake. Their once crude village became a huge, magnificent city and the rafts, invented in a gamble to stave off poverty, proliferated to keep pace with the demands of the capital city of central Mexico.

The sight of these islands astonished the conquering Spaniards. Indeed, the spectacle of an entire grove of trees seemingly suspended on the water must have been perplexing, even frightening. William Prescott, the historian who chronicled the destruction of the Aztec empire by the Spaniards, described the *chinampas* as "wandering islands of verdure teeming with flowers and vegetables and moving like rafts over the water." *Chinampas* continued in use on the lake well into the nineteenth century, though in greatly diminished numbers.

While it might be tempting to identify Central America as the source of practices similar to, but preceding, the development of hydroponics, no firm statement can be made. Other people living in similar conditions might have been equally ingenious in discovering the methods of "water gardening." While the *chinampas* of the Aztecs are not precisely hydroponics, they do use many of the same principles—an understanding of the ability of roots to exist in a water environment and the use of an altered, improved version of soil, although in their case, the improvements had already been accomplished by nature. At least one gardening writer has suggested that the Hanging Gardens of Babylon were in fact an elaborate hydroponic system, into which fresh water rich in oxygen and nutrients was regularly pumped.

The point of this brief detour into history is humility. Again and again, when we consider the long role of agriculture in human existence, we discover that our ancestors anticipated us. Too often, we have assumed that agriculture before the twentieth century was a primitive business, and its practitioners a dull, conservative lot. Not so. Agriculture has usually been distinguished by innovation, insight, and the desire for improvement. And this was not the work of cliques of scientists dictating to an awed and obedient audience. The history of agriculture is the history of common men and women, working, learning whatever their gardens could teach them.

You might want to keep this in mind when you start a hydroponic garden: despite the vast amount of information we already have on the subject, there is much left to be learned, many surprises, many as-yet-unexplored potentials. Any such possibilities can best be explored by numbers of gardeners working with practical results in mind. In a sense, because of its low cost and relatively easy workload, hydroponics has the potential to give gardening back to the gardeners, to make everyone a pioneer in the field.

PLANT NUTRITION

We have already mentioned the nutrients from which a plant draws nourishment. What are they, and how does a plant use them?

The twin processes of decay and erosion are responsible for depositing a wide variety of elements in the soil. When any living thing dies and falls back upon the surface of the ground, various bacteria go to work to break down its complex structure and derive nourishment from it. The bacteria take only a few kinds of elements—many more are released and thus returned to the soil. Inorganic materials such as rocks (and more recently man-made metals) are worn down, eroded by the actions of wind, rain, temperature and waves, reduced to tiny particles of elements that are deposited in the soil.

The elements we are concerned with are referred to as mineral salts. These elements—among them nitrogen, phosphorous, potassium, calcium, iron, copper, zinc, boron, manganese, and magnesium—are soluble in water. When water flows through the soil, the elements dissolve into it and are carried along. When plants absorb water, they are bound to be absorbing some mineral salts. However, the amount they absorb will vary widely, depending upon what has been deposited in the soil. Plants in the wild can compensate somewhat for a deficiency of these mineral salts, but they cannot do so indefinitely. Areas in which little is naturally added to the soil soon become desolate, having little plant life. That which they do have will be composed exclusively of those forms that have mastered the art of survival under the harshest conditions.

In his book *Plant and Planet,* Anthony Huxley has referred to chlorophyll as the "great invention of the plant kingdom." It is a substance found only in plants, where it plays the essential part in energy production. Unlike all other living things, plants do not simply consume material and extract energy from it. They are incapable of doing so. Instead, plants have evolved an entirely different method for assuring themselves of the "food" on which their growth is based. They manufacture it, they make their own. This process by which a plant produces its own food is called photosynthesis, and chlorophyll is the central element in the process. Without chlorophyll, it cannot be done.

Now we can begin to understand the importance of the various experiments described earlier. Joseph Priestley and Jean Ingen-Housz proved that plants absorb carbon dioxide, taking it in through the porelike openings called stomata located on the upper pads of a plant. What they didn't know was *why* plants needed the carbon dioxide.

Chlorophyll absorbs the rays of sunlight, and uses them in combination with water and carbon dioxide to generate glucose, the substance that fuels the growth of a plant. Carbon dioxide is an essential ingredient in the production of glucose. Oxygen is given off by a plant as a by-product of this process.

Woodward's work was only partially correct. Contrary to his belief, plants *don't* feed from the soil, but they do depend on the nutrients, the mineral salts, being held in it. And while he was right in maintaining that plants don't feed from water, it *is* the means by which the necessary mineral salts can be carried into a plant. Sunlight and carbon dioxide are absorbed in other ways, so only the mineral salts and water must be drawn into the plant. In reality, neither soil nor water sustain a plant. It sustains and nourishes itself, using what it absorbs as components of the manufacturing process it directs.

When the sun goes down, photosynthesis ceases. The energy generated throughout the day, in the form of glucose, is stored until the manufacturing process ends. Then, during the night, the glucose is released and put to work. After the sun has set, plants begin drawing in oxygen and giving off carbon dioxide—an exact reversal of the daytime process. Oxygen now acts as the triggering mechanism, causing the plant cells to release the energy manufactured during the day. This energy is used in the generation of new cells which extend the roots, increase the length of the stem and branches, and produce new leaves. Mineral salts are needed not only in the production of energy; they are also essential to the process by which new growth is produced and to the regular functions of the plant.

We have already noted that mineral salts dissolved in water are drawn in through a plant's roots. But how exactly is this done? By a process known as osmosis.

Projecting from the surface of roots are great numbers of bristly "hairs." These hairs are porous and actually draw in water. Roots keep on growing because they are constantly exhausting the supply of water in their perimeter and must continually push out to tap new supplies. Following the pull of the earth's gravity, roots grow down. However, following the need for water, they also grow horizontally. The conflict between these forces causes roots to grow using a spiral movement—roots rarely run very far in a straight line. The root hairs occur just behind the tip of a root. As the root grows, the old hairs die and new ones appear close to the extended tip. The hairs actually have the capacity to grip particles of soil, as you may notice when lifting a plant free from the earth. Soil particles appear to be clinging to the roots. In fact, the roots are just as much clinging to them.

Water is transferred upward by way of the xylem, a pipelike structure of cells running vertically through the plant. Energy is conducted downward along a column of cells known as the phloem. Mineral salts are also passed along through the phloem.

I have gone into some detail on the matter of plant nutrition for several reasons. First of all, I want to stress the idea that plants are remarkably complex creations, self-contained survival machines making whatever they need to endure and to successfully reproduce. While we admire plants, too often we dwell only on their appearance, unaware of the complexity of the modest creation before us. Plants are a terrific evolutionary success, less dependent on the world around them than most other creatures.

Knowing about plant nutrition will also help you understand something of the *why* behind the success of hydroponics. The hydroponic method works because it makes such perfect use of a plant's systems and needs. Soil gives plants support and serves as the reservoir of mineral salts and water. Hydroponics allows you to supplant both these functions, and if you do it properly a plant will not only adapt to a soilless existence but actually become healthier and more productive as a result.

HYDROPONICS AND MODERN AGRICULTURE

Let us grant the fact that a hydroponic system works, and works efficiently, cheaply. Let us even say that it is ideally suited for use in urban areas, where few people have access to plots of land. What makes it more than a good but unimportant addition to conventional agricultural methods?

Those very agricultural methods make hydroponics of such great importance. The Western world is caught in an energy shortage that seemingly can have only one outcome—the utter depletion of the fossil fuels our systems depend on. It may be that the development of nuclear and solar energy systems will partially remedy the problem, but such solutions will have little effect in agriculture. *Every day,* two hundred thousand barrels of oil are used to manufacture nitrogen fertilizer—over a million barrels of oil *a week* just to keep pace with the current demand. For every unit of food produced, 6.4 units of energy must be expended. By the time that food has been processed, as much as fifteen units of energy may have been consumed. In some cases, it has been estimated that the tractors used to prepare a field burn up as much energy as the entire field can yield.

We have to begin using energy wisely. And hydroponics can be a very smart way to grow plants, since nutrients can be used over again, recycled through the system, thus greatly cutting down the amount of fertilizer that must be applied to produce a crop.

There are other difficulties inherent in modern agricultural practices, most notable the trend towards very large farms that stretch for thousands of acres. Contrary to the line usually advanced in behalf of such farms, they are *not* more efficient than small farms, those worked by the owner and his family. USDA figures indicate that small farms actually outproduce large farms. There is, however, one thing that large farms accomplish—they make money for the corporations that own them. More and more land is being gathered up by corporations and intensively worked by large machines. The need for workers is reduced to a minimum by the application of immense amounts of fertilizer and other chemicals. In the short run, heavy mechanization and the production of just one or two crops over a very large acreage do increase yields and profits. But in the long run, such practices are dangerous, destructive, and perhaps permanently damaging to both the land and the economy.

Immense amounts of energy must be used to make these maxifarms profitable. Prodigious quantities of chemicals must be continuously pumped into the farms to maintain crop yields. Because little attention is paid to the quality of the soil, to thoroughly aerating it, to improving its structure by working in organic materials, the only way that crops can be regularly produced is to pack the soil with fertilizers. Soil is a remarkably complex natural structure with an intricate balance of components. It is not simply an automaton, a machine which will then unfailingly produce the necessary results.

When you treat it as nothing more than the container of a crop, you are bound to unsettle it badly. And then you must pump in the fertilizer just to keep crops going. However, each year, as the soil further deteriorates, you need more fertilizer to accomplish the same purpose. In some mechanical farms, the rate of application of fertilizer has climbed as high as an incredible five hundred pounds per acre. As the soil deteriorates, as much as fifty percent of this fertilizer is simply leached away, never reaching the crop.

A similar process has been followed in the development of high-yield plant varieties. Strains have been produced that give much greater yields—but at the cost of immense amounts of chemicals. In an article on "The Withering Green Revolution," Marvin Harris has noted, "The main problem with the miracle seeds is that they are engineered to outperform native varieties only under the most favorable ecological conditions and with the aid of enormous amounts of industrial fertilizers, pesticides, insecticides, fungicides, irrigation and other technical inputs. Without such inputs, the high-yielding varieties perform no better—and sometimes worse—than the native varieties. . .especially under adverse soil and weather conditions. . . . Even when the technical inputs are applied in sufficient quantities, certain ecological problems arise, which seem not to have been given adequate consideration before the seeds were 'pushed' out into the vast acreage they now occupy. Conversion to high yield varieties creates novel opportunities for plant pathogens, pests, and insects. The varieties also place unprecedented stress upon water resources."

And there are other problems. The use of heavy machines badly compacts the soil. Much of the land used by "agribusiness" corporations is left bare, without a cover crop, during the winter. The already damaged soil is especially vulnerable to the workings of the wind and rain, and is carried, washed, and blown away in large quantities. In the United States alone it is estimated that *three billion* tons of topsoil are lost to erosion every year, the greater part of this loss being the result of the actions of humans.

When fertilizer is lost from the soil, where does it go? Usually into the nearest water system, carried there by the flow of ground water. Once in a stream, lake, or river, it can accumulate to dangerously high levels and cause the wild proliferation of algae and other similar plants. These rapidly grow into such great masses that the local life forms feeding on them cannot control their spread. As these masses decay, they release quantities of bacteria. In breaking down the plants, these bacteria use oxygen, the loss of which serves to kill off other life forms depending on it. Eventually, the body of water will be largely lifeless and the masses of plants, transpiring water from the lake into the air, will cause the level of water to steadily decrease. A swamp exists where there was once free-flowing water.

Residues of fertilizers can also end up in domestic water systems. Most water treatment plants are not equipped to filter out plant nutrients since it requires a long and expensive process to do so. While small amounts of chemical fertilizers may not be harmful, large doses can be poisonous. Fertilizer nitrates can contaminate water supplies, making them unusable.

There have already been indications that conventional agribusiness farming is in trouble. Land that as little as twenty years ago was among the most fertile has begun to produce smaller and smaller yields. Another even more frightening sign of trouble is the proliferation of new strains of pests and diseases resistant to pesticides and other chemical treatments. Insects reproduce with such frequency that complete resistance to a poison can be developed in only a few years. On several occasions, pesticides have proved more destructive to the natural predators of plant pests than to the pests themselves.

While all of these problems will have an effect on us at some point in the future, several related problems are even now causing us difficulty. It already costs as much to transport food as it does to grow it. In many cases, it costs more. And food can only become more expensive.

In their book *Robots Behind the Plow,* Michael Allaby and Floyd Allen predict that "Farming on a large scale, using large numbers of large machines and heavy applications of artificial fertilizers and pesticides, will be seen to be an expensive way to farm. . . . When to the production costs are added the costs of transportation, storage, processing, packing and handling, the difference in cost between the large-scale production and distributor operation and the small-scale farmer, using fewer machines, no artificial chemical products, and selling locally, will become very evident. The consumer will receive less and less food for every dollar spent." They conclude their analysis of the failure of big agriculture, or corporate farming, by observing that "We expend perhaps a hundred thousand times more energy producing the food and bringing it to you than the food will yield. So long as we are able to believe that this energy is provided almost free of charge, the illusion will hold. It will not hold for much longer" (p. 57).

There is an additional effect the consumer must already reckon with. The requirements large-scale growers apply to a crop are not necessarily the same as those a consumer embraces. The tomato provides the most pointed illustration of this difficulty. Large-scale farmers have almost unanimously come to rely on a few varieties of tomatoes. These varieties are used because an entire crop can be brought to maturity at the same time and thus harvested in a way best suited to the use of much machinery. These varieties have a regular shape so they can be easily fitted into plastic containers. They also feature thick, tough skins so they can better withstand rough handling during shipment. There is only one quality lacking in such varieties—flavor. In some cases, tomatoes are picked while they are still quite green, held in cold storage until they can be distributed, and then gassed to give them the red color we associate with freshly picked ripe tomatoes. And tomatoes are only one example of current practices within the agricultural industry and the effect they have on the food we eat. Every mass-produced fruit and vegetable is affected in some similar way.

But what does all this have to do with hydroponics? Potentially, quite a lot. Growing plants hydroponically—whether they be house plants, vegetables, or herbs—makes much better use of such limited resources as fertilizer and water. Even a small hydroponic unit will produce sufficient vegetables to provide a regular supplement to a family's diet of store-bought goods—vegetables free of pesticides, and naturally sweet and ripe. If you have no land, a hydroponic unit allows you to declare your independence, even if only in a small way, from the practices of agribusiness.

2. Methods of Hydroponic Gardening

Hydroponics is an umbrellalike term covering a wide variety of methods. The basic principles of hydroponics can be applied in a great many ways, fitting themselves to the financial requirements and space limitations of each gardener. The methods of hydroponic gardening developed over the past fifty years can be divided into categories defined by the medium in which the plants are grown. In *water culture,* the plants are grown only in water, or in a water and nutrient solution. *Sand culture* calls for raising plants in a sterile sand, into which a water and nutrient solution is pumped. *Aggregate culture* replaces the sand, using instead any one of a number of materials such as gravel or vermiculite, and retaining the method of pumping water and nutrient solution into the material. Finally, there are a number of experimental or unorthodox techniques.

BASIC WATER CULTURE

Growing plants in a container filled with a water and nutrient solution is the simplest and least expensive method of hydroponic culture, thus making it the ideal introduction to the science of hydroponics. Water culture is best suited for containers in which only one or two plants are displayed, limiting the profitability but not the attractiveness of this method. If you've never worked with hydroponics before, this is the best way to gain experience with the principles involved, even if you intend to move on to more complex practices.

The Container

Most of the commonest house plants do very well in a water and nutrient solution. Indeed, even some kinds of cacti will prosper in water, making a startlingly different display. Later, we'll discuss some of these plants individually, but first we should establish the basic procedures. Growing house plants in water is a well-established practice in Europe, and manufacturers have produced a variety of handsome glass containers to meet the demand. Unfortunately, few containers designed specifically for such a purpose are available in America, so it is still necessary to use one's ingenuity. Actually, anything that is waterproof and of the appropriate size will do. I prefer clear glass containers because they allow an unobstructed view of the root system. One of the best reasons for growing plants in water is the view one is afforded of the roots—they become a part of the aesthetic appeal of the plant, along with the leaves and blooms.

Despite the attractions of glass, your choice of a container is visually unrestricted. Any sort of vase, bowl, pottery, or bottle may be put to use. You can achieve some unusual effects by recycling bottles, especially those sorts not usually thought of as containers for plants. I know of at least one gardener who has adapted a variety of wine bottles, each distinguished by its shape, texture or color, and arranged them to form an original and pleasing display, placing in each bottle a plant that agrees or contrasts with the container's color or texture. The glassware used in laboratories is inexpensive, available in an astonishingly wide variety of sizes and shapes, and is distinguished by its simplicity of design and purity of form.

Some plants require more support than suspension in water will give. For these, you must cover the bottom of the container with a layer of gravel or some other solid material. This allows the roots to find an initial degree of support. As the plant grows, the roots will proliferate until they spread throughout the container and are in effect self-supporting. The gravel sold in pet shops for aquariums is excellent for this purpose. Glass marbles are also acceptable, and sometimes more attractive. Small shards of pottery or

Containers for growing plants.

pieces of brick will also serve, but you must scrub them thoroughly in hot, soapy water.

Just as any material placed in a container must first be sterilized, so must the container itself. Chemical supply houses selling glassware designed for laboratory use also generally sell brushes with very stiff bristles. Such brushes, used in combination with hot soapy water, are sufficient to cleanse any of the materials you'll be using.

Allowing a plant to lean against the side of a container should not prove damaging to the plant. However, it may give the plant a rather careless or untidy appearance, as if you'd simply thrust it into the water and forgotten about it. If this disturbs you, it can be remedied quite easily. Use only those containers that come with cork stoppers, or for which you can supply a cork stopper. Carefully drill a hole through the center of the cork, making the aperture large enough for the plant to be passed through. After you've gotten the plant through the cork, you can support its stem in place by packing the opening with sterile absorbent cotton. This method was successfully employed as early as 1865.

The Water

You can use tap water to fill the containers in which you place plants, but you'll have to check the water regularly to determine its pH level. A variety

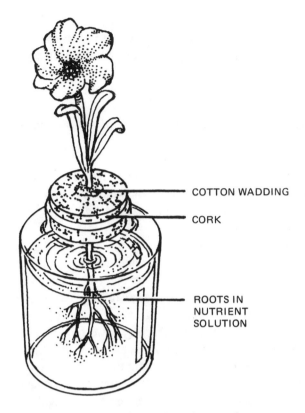

Growing a plant in water, using cork and cotton for support.

of inexpensive pH testing devices are available, both from seed firms and at some garden supply centers. (See the list of suppliers in the Appendix.)

Any wet or soluble substance has the capacity to retain some amount of the element hydrogen. This amount can be determined by testing and is represented in a numerical reading along a scale ranging from 1 upwards. A high level of hydrogen turns a substance alkaline. A low level of hydrogen makes a substance acidic. A reading of pH 7 is considered the center of the hydrogen scale—that point at which the alkaline and acidic qualities are in balance. Any reading above pH 7 indicates that the substance is alkaline, while a reading below pH 7 means that the substance is acidic. But what does all of this have to do with plants?

While the matter is complex, what we need to know is this— house plants prefer a growing medium, whether water or soil, having a pH reading between 6 and 7. House plants can draw in nourishment through their roots *only* when the pH level falls within this range. If it is much above or below this range, the plant will be unable to absorb the necessary nutrients and will lack the ability to continue to fuel its growth.

The likelihood of having an unacceptable pH level has been greatly increased by the variety of chemicals now used in water treatment, chlorine being foremost among the culprits. Chlorine will disappear from water that is allowed to stand for two days without being used. If your water tests out too alkaline (as is usually the case) pour it into a container and begin adding drops of vinegar. Keep testing. When the pH level has been brought within the proper range, record how many drops of vinegar it took. Each time you intend to replace the water in your garden, fill up large containers with water, test it, and add vinegar until you get a reading of pH 6 to pH 7. Tap water is rarely too acidic, but if yours should test out so, follow the treatment above, substituting bicarbonate of soda for the vinegar (see section on pH, p. 61).

Never use water that has passed through a home water softener. Water so treated is inimical to plants.

While some house plants will grow prolifically in water alone, some positively need and all will benefit by the addition of a soluble plant food. We'll discuss nutrient solutions at length later in the book, but for our present purpose almost any general plant food available in your area will be adequate to your needs, so long as it can be dissolved in water. Powders are the easiest to measure out for such uses. As a very powerful dose of nutrients is potentially more damaging than a dose that is too weak, it is a good idea to reduce the amount recommended by the manufacturer by one third.

Caring for your plants can be greatly simplified by changing the water in containers on a regular basis. If you have many, do half or a quarter of the containers at a time. If there is much chlorine in your water, you'd best begin by drawing the water two days in advance. Test it to determine the pH level the day before you intend to use it, and make any alterations in the pH that are necessary. Just before you intend to use the water, mix in the nutrients. If you have plants growing in many containers, it would be easiest to prepare the water and nutrient solution in large batches. A bathtub is suitable, or you might want to keep several gallon-sized containers on hand and reserve them for this purpose.

The water in each vase or other container should be changed once every three to four weeks. Plants are not the only living things that do well in a water and nutrient solution—so does a wide variety of algae. Changing the water no later than every month should avoid any real problem. However, if masses of green muck appear on the water before a month has elapsed, change the water immediately, scrub out the container, and cut down on the strength of the nutrient solution you've been using. Algae is *not* harmful, just unsightly.

In the weeks between each change, do not add additional nutrients. As the water in a container evaporates, simply add more water. No matter how tempting it may be to add just a dollop of nutrients to each container, avoid doing so. The strength of the solution would be increased greatly, perhaps to the point of damaging the roots.

The Plant

You can place either cuttings or an entire plant in water. Cutting can be taken from a plant and immediately placed in a container filled with water, but before you transfer a mature plant you'll have to thoroughly, gently clean all the soil from its roots.

Spread newspapers across a table or any other flat work area. Place one hand over the soil surface of the plant, positioning the stem securely between two fingers. Grasp the base of the pot with your free hand, invert the pot and carefully pull out the plant and its root ball. If the plant will not come away, tap the pot carefully several times against a hard surface. If the plant still will not slide out, run a dull knife around the perimeter of the soil.

When you've got the plant out, hold it by the stem—but don't pinch it or grip it tightly. Use your free hand to slowly, gently brush away all the clots of soil still clinging to the plant's roots. You may then either hold the roots under a mild flow of lukewarm water or stand the plant in a shallow container of lukewarm water so that the roots are entirely immersed. You may want to try either procedure or both—the point is to touch the roots as little as possible. They are easily bruised, and a bruised portion can quickly develop rot. If the roots fail, the plant dies.

Don't add any nutrient solution when first transferring a cutting or plant to water. After a week has elapsed, pour out the water and pour in a water and nutrient mix.

As mentioned earlier, some plants at first require a degree of support in the form of a layer of pebbles, marbles or shards of crockery. Even plants not requiring such support will benefit from having several pieces of charcoal on the bottom of the container. The charcoal, available at plant shops and garden centers, acts as a natural purifier, serving to keep the water clean and pure, and to retard somewhat the growth of algae. Drop the charcoal into each container, then add a layer of some coarse material, if it is called for. Next, stand the plant in the vessel and slowly pour water around it until the roots and at least part of the stem are under water. Don't fill the container so full that leaves become sodden or are covered by the water—they will quickly rot. If the plant is to gain support from pebbles or some other material, its roots must be gently spread over the surface of the material. Some of the thicker strands of roots can be covered by the material, so long as a very large pile is not heaped over them.

Caring for the Plants

Once a plant has adapted to its watery existence, little care is required. Any possibility of soil-borne insects or diseases is immediately eliminated. While white fly, spider mites and aphids may still settle on foliage, their presence is made less likely when you do away with soil. If they do occur, they will be very visible, and can be destroyed if treated promptly.

The most difficult thing you will have to do is to keep the plants' leaves free of dust. Even in the cleanest rooms, there is always dust in the air. And if it's there some of it is bound to end up on a plant's leaves. You can't avoid it, but you can remove it if you work on a regular basis, checking each of the plants at least twice a week. A sponge dipped in lukewarm water or a clean, soft cloth can be used to take the dust off.

Levels of pollution are so high in some areas that plants will soon acquire a visible layer of gritty particles. These must be removed from leaves on a regular basis, and not simply for aesthetic reasons. Dust accumulating on a leaf will eventually clog its pores and thus interfere with the vital process of photosynthesis, during which a plant generates the energy to sustain its growth. If a plant becomes very grimy, remove it from the container and gently wash the leaves and stem with lukewarm soapy water. Use your fingers to massage the plant surfaces with the soap. Rinse the plant under a mild stream of cool (not cold) water.

There is a slight chance that the roots of a plant displayed in a container of clear glass could be burned if the container is placed in a spot receiving many hours of intense, direct sunlight. The roots don't need light, only the leaves do. If you have such a spot in mind for a plant, try to position the container so that the sunlight falls only on that part of the plant above water.

Plants in water are less subject to damage by inadequate amounts of humidity than plants grown in soil. In fact, each plant's container acts as a ceaseless, natural humidifier, moistening the air as the water evaporates. On very warm, dry summer days, you might want to supplement this source by gently misting the leaves of each plant.

ALTERNATE METHODS OF WATER CULTURE

There are several other methods of water culture. All of them work, none of them are particularly expensive to set up, but they are all more elaborate than the method described above.

The *continuous flow method,* for instance, calls for the use of three containers. These must be arranged so that they occupy three different levels, one above the other. The highest container holds nutrient solution. From it a pipe or piece of tubing extends into the middle container, where the plant is growing. Another piece of pipe or tubing must extend from near the bottom of the middle container, carrying the nutrient solution down to the third and lowest container, where the solution is collected. In this way, a continuous flow of solution is established. As the topmost container empties, it can be refilled from the lowest container.

I can imagine such a system being quite impressive, a fascinating sight. But it is not very practical. It would take a good deal of testing to reach the right angle for the pipes, so that the solution would flow neither too quickly nor too slowly through them. The containers and pipes could be purchased at a store carrying scientific supplies, but it would probably take some searching to collect everything you needed.

While such a unit might make an impressive display, it is still rather impractical. All this for just one or two plants? Another difficulty in adapting this system to home use is the necessity of regularly aerating the solution by pumping oxygen through it. The system actually functions best when applied on a much larger scale for experimental or commercial purposes.

The *Swiss method,* also known as "Plantanova," calls for an ovoid (egg-shaped) vessel. The top quarter of the vessel lifts off, allowing easy access. Inside, positioned at the midpoint of the container, there is a small tray which can also be lifted out. The area below the tray is filled with nutrient solution. The tray itself is covered with a layer of stone chips which position and support a plant. Its roots grow down into the solution, and its stem passes upwards through a hole in the center of the vessel's dome. Such containers are not yet generally available outside Europe.

The *Gericke method* is named after the man who perfected it, and thus allows you to duplicate the system that first brought hydroponics to the attention of the public. The method calls for the use of waterproof tanks or basins. These containers should be long, but they need not be deep. A wire grid is fitted over the containers, and over the grid a mixture of peat and sawdust (or hay) is spread. Plants are placed in this material so that their roots may grow through the grid and their stems above it. The nutrient solution is poured in before the grid has been dropped in place. There should be a space between the grid and the nutrient solution so that the roots growing down towards the water will also receive oxygen.

Besides keeping the plants upright, the mix of hay and peat moss keeps sunlight from falling on the solution. Sunlight could be damaging to the balance of the nutrients in the solution, but it is absolutely necessary for the plants and is needed in abundant amounts. Indeed, the Gericke method has been used most successfully on a large scale in areas receiving intense sunlight. You could implement a Gericke unit on a small scale in your home, but you would have to supplement the natural light with an artificial light unit.

HYDROPONICS IN FLOWERPOTS

You can also raise plants in flowerpots filled with any growing medium. In fact, if you have started out your experiments with hydroponics by raising plants in containers filled with water, moving on to flowerpots filled with

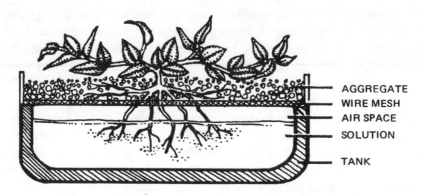

AGGREGATE
WIRE MESH
AIR SPACE
SOLUTION
TANK

Growing a plant in a tank, with aggregate for support.

a medium is the next logical step. It's a bit more involved, but it will further advance your knowledge of hydroponics and increase your confidence. After learning about the use of growing mediums by experimenting with them in flowerpots, you'll be ready to move on to running a hydroponic unit containing many plants. Of course, you may want to continue using the flowerpots, for such containers will fit into spots where a larger unit could not be placed. By using a variety of different-sized containers for your plants, you can make the best possible use of every inch of space.

A pot is prepared for hydroponic growing in much the same way as if you were planning to fill it with soil. Pieces of crockery must be laid over the drainage hole to prevent the growing medium from draining out. Although any sterile growing medium can be used to fill the pot, gravel seems to give the best results. Fill the bottom third of the pot with large pieces of sterile gravel and use smaller, finer pieces to fill the remainder. When the pot is almost two-thirds filled, gently stand the plant in place. If the plant has been removed from soil, all traces of it should be removed from the roots. Hold the plant under a gentle stream of water and gently remove any clots of soil that remain attached. You should touch the roots as little as possible. While holding the plant upright, slowly brush in sufficient gravel to cover the roots and keep pouring gravel in until you have reached the plant's soil line—that point at which the stem of the plant emerges from the soil. The portion of the stem below the soil line is darker than that greater part of the stem visible above the surface and so can be easily recognized.

After you have properly arranged the plant, pour half a glass of water onto the gravel. Wait for fifteen minutes, then check to make certain the water has poured unobstructed from the drainage hole. It's important that the crockery is arched over the hole, serving to retain the gravel while allowing the water to flow through unobstructed. If you prefer not to use crockery, a piece of screening having a very close mesh could be laid over the drainage hole, and will serve the purpose well.

SAND CULTURE

If you want to grow a number of plants in a single container, if you want to raise plants in a greenhouse or in a large-scale hydroponic unit, you'll have to go beyond simple water culture and use a soilless medium to give the plants support. Sand has been a frequent choice since the thirties. It is sterile, retains moisture well, and can be used with equally good results in both large and small units.

However, using a soilless medium introduces a further complication. Since the plants are no longer growing in a water and nutrient solution, this solution must be delivered to the roots in some other way. A number of procedures for this have been worked out. There is, for instance, the surface watering method, which works much the way it sounds. The water and nutrient solution is applied to the surface of the sand by a watering can or poured from a jar. The liquid gradually soaks through to the roots. While this process works, it is also rather wasteful since some of the solution evaporates. In addition, a unit receiving solution in this manner must be watered several times a week.

The *New Jersey methods* (also known as the Withrow methods) call for setting up separate reservoirs of solution which are applied to sand on a regular basis. In one version a container is set up above the unit, and solution flows from it through a hose down to the sand. Thus the influence of gravity is utilized to deliver the solution. In another version, solution is introduced into a unit by means of a small pump. Of course, waterproof containers are required. An additional feature of such a method is the ease with which the pump can be automated, timed to release the solution on a schedule.

I have said that it is impractical to try to grow numbers of plants in a single container filled with water alone. As mentioned in the previous section, plants will do well in water if they are grown in individual containers, or in specially designed units. You can also grow single plants in sand. While the method works, it seems to me to be too costly to be practical.

The process involves a flowerpot, filled with sand, in which the plant is placed. A wick fashioned from cloth or nylon runs from the drainage hole of the pot down into a container holding nutrient solution. The wick, which extends at least an inch upwards into the sand filling the flowerpot, acts to steadily draw solution up from the container and into the pot. The end of the wick inside the flowerpot should have ends fanned out through the sand so that the solution will be equally distributed to all of the plant's root system. It will take some experimentation to get the wick working properly, but it *will* work. You may enjoy the unique nature of the display, but it is obviously not practical for growing large numbers of plants.

Other methods for applying solution to sand exist, but most are either too elaborate or still in the experimental stages of development. Since I have discussed the more elaborate methods of watering, I suppose I should also mention the simplest. All you need do is sprinkle the surface of the sand with fertilizer salts and pour water over the surface. The water dissolves the salts, the sand absorbs the water, and the roots eventually absorb the nutrients. While this method works, it seems to me to be very inexact. The nutrients may not be distributed equally to all of the plants. There may not be

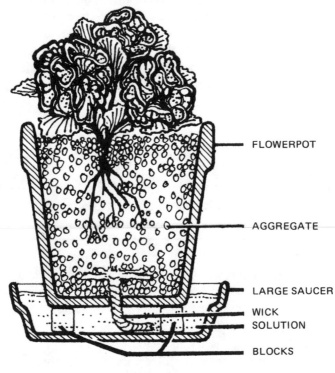

FLOWERPOT

AGGREGATE

LARGE SAUCER
WICK
SOLUTION
BLOCKS

Wick-watering method.

sufficient salts applied or, just as harmful, one may apply too much. An excessive concentration of fertilizer salts can seriously damage a plant's roots. It may be possible to monitor the supply of salts and make this system work well in a laboratory setting, but I think it is too difficult a method for the home gardener, demanding as it does a degree of very sophisticated control. Upon closer inspection, the simplest method turns out to be not so simple after all.

AGGREGATE CULTURE

While sand is generally a trouble-free medium, some hydroponic gardeners claim that it has a tendency to become waterlogged and is rather wasteful of nutrients. I don't think any final decision can be made on such claims, but if you want to try another soilless medium you have plenty of choices. Gravel is probably the single most popular soilless medium, used with even greater frequency than sand, but there are many others. Rockwool, perlite, vermiculite, shards of brick, sawdust and wood chips have all been used with excellent results. I have even seen large heads of lettuce sprouting incongruously from a sheet of styrofoam, into which a nutrient solution had been regularly pumped.

Gravel

Gravel is an excellent choice for anyone beginning a home hydroponic unit. It's easy to keep clean, cannot become waterlogged, and, while heavy, can be handled without difficulty. It's also inexpensive. However, when used alone it has the drawback of drying out quickly, necessitating frequent applications of water. You can circumvent this problem by mixing sand and gravel together to form your medium. The sand will act to retain moisture, while the gravel will prevent the sand from becoming sodden and waterlogged. The longer the period you desire to have between waterings, the more sand you'll have to add. A mix containing five parts of gravel to three parts of sand should prove workable in most instances. Such a mix will need to be watered once a day. By increasing the amount of sand in the mix you can lengthen the periods between waterings. No matter what the percentages are, the sand and gravel must be thoroughly mixed together. A section of the unit in which either material predominates will throw the entire unit out of kilter, causing some plants to receive more solution than they can absorb and others less than they need.

As with other aspects of this still-developing field of hydroponics, you'll have to do some experimenting to determine what works best for you. Use the ratio of 5 to 3 as your starting point. If it doesn't work, experiment by adding more sand, or by adding a finer or coarser grade of sand, or by adding more gravel.

Rockwool

Rockwool is a recent addition to the choices of growing mediums. This sterile, porous, non-degradable material is composed primarily of limestone or granite which has been subjected to intense heat and spun into long fibrous strands. Rockwool can be molded into blocks, sheets, cubes, or even crumbled into granules. It absorbs moisture without absorbing nutrients, which means that all of the nutrient solution will get to the roots and not be locked up in the growing medium. Rockwool can support a number of plants in a hydroponic unit; or a slab of rockwool can support its own individual seedling with water and nutrient solution supplied through tubes connected to a central reservoir.

Like every other growing medium, rockwool has its drawbacks. When it's dry, it is rough enough to irritate your skin. When you're preparing rockwool, you'll have to wear gloves. I'd recommend goggles as well, to keep the dust of dry rockwool out of your eyes. Once rockwool is wet, it's safe and very easy to work with. Because rockwool retains moisture easily, algae can quickly appear on its surface. Algae growth won't affect the health of your plants, but it looks unpleasant and may attract a variety of small flying insects. To prevent it, many growers buy rockwool packaged in square slabs, covered in plastic except for a round opening large enough to admit a seedling. If you want to use one large slab of rockwool in a hydroponic unit, cut regularly spaced holes in a sheet of thin, dark plastic and suspend it just above the surface of the rockwool.

Perlite and Vermiculite

Perlite and vermiculite are easy to work with. Both are primarily composed of minerals, and their heavy particles expand when subjected to heat to become absorbent and very light.

Because perlite and vermiculite are so light, they are ideal for areas where a heavier material would present problems, such as rooftop units or units suspended from the ceiling. They are, in my opinion, the best choices for any apartment gardeners. Both perlite and vermiculite are attractive materials; if you're gardening in your apartment or home, rather than outdoors or in a greenhouse, the handsome texture of the materials attractively complement your thriving plants.

Perlite may be used without the addition of other materials in a unit. It retains water but is poorly permeated by minerals, which means that it won't lock up nutrients intended for your plants. I've always had very good luck with perlite. One note of caution, however: when you first open a bag of perlite, do so outdoors. The fine dust in the bag can have a high silica content, which can irritate your lungs. Pour the perlite into an open container and direct a fine spray of water onto the material. Transfer the freshly washed perlite to your unit or a storage container.

Unlike perlite, vermiculite can become waterlogged very rapidly; so, it must be mixed with a coarse grade of sand (at a ratio of two parts vermiculite to one part sand) or perlite (at a ratio of one part vermiculite to two parts perlite) to allow water to flow evenly throughout the hydroponic unit.

Shards of Brick

Shards of brick can be used as a soilless medium if they are small enough. They will work, but they are not practical. They're difficult to work with, must be thoroughly scrubbed before they can be used, and the material from which they are formed may begin to decompose and drastically alter the pH level of the solution. I suppose using pieces of brick as a medium in which to raise plants provides a powerful demonstration of the adaptability of the hydroponic method. But that's not why you want to grow plants hydroponically.

Other Mediums

Sawdust can be used as a soilless medium, as can hay or straw. But sawdust has a tendency to clot, to stick to a plant's roots, and to become compacted by water. Straw and hay, being composed of organic materials, will eventually decay, disintegrate, and cause a prime mess while doing so. In addition, they may harbor diseases or pests deadly to your plants.

I think styrofoam serves as an excellent example of a tendency in hydroponics that can mislead and play havoc with the novice. Yes, you can grow plants in sawdust, in hay, in brick shards, in styrofoam—but that doesn't mean you should. To do so provides excellent examples of how remarkable the hydroponic method is—and providing such examples is important for those who doubt that the system works. But *you* don't need that proven to you. Leave it to others.

For the home gardener, it is more important that the method employed works with as few problems and expenses as possible. Every growing medium has its own advantages and drawbacks. As with every other aspect of gardening, you'll find gardeners who swear by one medium and damn all others. The best thing to do is to experiment. Try several to see which one works best in your own environment.

Many manufacturers and suppliers offer a variety of hydroponic kits. These kits often are sold either with an initial supply of a growing medium, or a suggestion as to which medium to use. Start by following the manufacturer's suggestion—these units are designed to function best with one particular medium, or a specific mix of several, but they are not designed to function with just any medium.

Irrigation

As with sand culture, a number of methods of irrigation have been worked out specifically for aggregate culture.

The *sub-irrigation method* calls for a watertight container into which a diluted nutrient solution is regularly pumped, flooding the unit. Then the container is allowed to drain. The system is automated and is said to be especially efficient. However, it is somewhat costly to install. A complete sub-irrigation setup would be practical only for a large hydroponic unit, such as that established in a greenhouse. However, one could modify the plan and use the principles of sub-irrigation for even a very small unit.

The *bucket and gravity feed method* is reminiscent of the New Jersey sand culture methods. A container filled with nutrient solution is attached to a post standing above the hydroponic unit and connected to it by a hose. The hose runs from the base of the container to the bottom of one of the side walls of the unit holding the plants. The solution will thus run down through the

An example of the bucket and gravity method in a more permanent form than that described in the text. In this case the bucket containing the solution has been given a permanent position above the unit. A faucet at the opposite end, bottom, can be opened to allow the solution to drain out into a plastic pail. This pail can then be emptied into the container for the solution, and the process repeated. Note that the hose leading from the container of solution to the unit can be closed to prevent solution folowing into the unit when it is unwanted.

hose into the aggregate, reaching the roots. When all of the solution has run out, the container is lifted from its post and placed on the floor. Any excess solution will run back down from the unit into the container and may then be re-used. This method is inexpensive and simple but effective, making it a good choice for the home gardener with a small hydroponic unit.

The *compressed air technique* is interesting, but really feasible only for large-scale commercial units. It involves the use of air compressors to drive the solution up into the hydroponic units from containers located on the ground.

The *modified slop method* uses a pump to flood the surface of the soilless medium with water. Gradually, the solution filters through the aggregate, over the roots, and finally flows out through a drain which is connected to a container under the unit. The excess solution that collects here can then be re-applied to the unit. In such a closed system, very little solution is wasted.

While these irrigation systems bear a certain resemblance to the methods developed for use with sand, there are other even more elaborate methods of applying solution to aggregates. In fact, almost any system that thoroughly distributes the solution throughout a unit will do. You needn't go to great lengths or great expense to irrigate your unit. Do only what you must. You don't need an automatic irrigation system unless you're not going to be around much. You don't need a large, complex irrigation system if a simpler design will do the job as well.

My point in mentioning some of the possible irrigation methods is to indicate the variety of choices and techniques you have available to you. Irrigation is one aspect of hydroponics in which you can use your native powers of ingenuity to come up with a design of your own. And if you can, you should. Sort through the methods mentioned, and think about which best meet your needs and budget. But don't feel bound to use them—you are the best, the most responsible, judge of your gardening needs.

EXPERIMENTAL HYDROPONIC METHODS

New methods of hydroponic gardening are constantly being introduced. Many are still in the experimental stages. Some have already been adapted by companies or individuals and are in use in a variety of situations, both urban and rural. Many of these newer methods require more elaborate equipment and some previous knowledge of hydroponic techniques. They are mentioned here only to indicate the vital state of research on the subject, and to underline the adaptability of the hydroponic principle.

The *floating raft method* is a contemporary version of the oldest known hydroponic gardening technique, the Aztec *chinampas.* On these rafts, food crops and flowers would be cultivated, their roots working down through the soil piled on the rafts to spread out into the water. The contemporary *chinampa* uses lightweight modern materials and in some versions is supported by large inflated inner tubes. A nutrient solution is regularly applied to the aggregate in the raft and allowed to work its way down to the roots.

Several hydroponic systems have been developed to produce grass or other forage for livestock. These systems include series of trays, some method of maintaining steady temperatures, and in many cases automatic machinery for pumping nutrient solution into the trays. Complete self-contained units in which new crops of fodder can be generated, from seed to harvest, in about seven days are now being marketed to farmers.

The *germination net technique* is used only for seeds or for bean sprouts. A sheet of mosquito netting is clipped into a container of hot paraffin wax. Before the wax has had a chance to dry, the netting must be stretched across a tray of nutrient solution. After the wax has hardened, seeds or sprouts are scattered across the netting and then moistened. The roots of the sprouting seeds will seek out the nutrient solution, but will at all times be exposed to the air. The netting must be placed very close to the solution, but should not be touching it.

Ring culture is the term for a system developed in Europe. Basically a form of aggregate culture, it is suited for conditions outdoors in areas having a high relative humidity or receiving much rain. Plants are placed in pots or other containers that have been filled with a mixture of sand, sawdust and peat moss. The containers are in turn placed on a two- to three-inch-thick bed of coarse sand, gravel, cinders or crushed brick. Nutrient solution is regularly pumped into the bed, where capillary action serves to draw it up into the containers. In addition, many of the plants will begin to send their roots downward into the bed. Although it is impractical for use with food crops, if you would like to bring house plants outdoors for the summer this method seems to me to provide an excellent way for you to insure their receiving sufficient water.

Large plastic bags were successfully used as containers for aggregates in one series of experiments. These bags were filled with any one of a number of aggregates—sand, charcoal, straw, even styrofoam. The bags were laid on the ground, holes punched in their upper surfaces, and plants inserted through the holes. The idea is for the roots to spread out into the aggregate while the stem and leaves remain above ground level. Nutrient solution is poured into the bag in a sufficient quantity and at regular intervals so that the aggregate is kept constantly moist but does not become waterlogged. This method seems to me to be inapplicable to most home gardening situations, but it could be useful in an emergency when you lack the materials or the time to immediately preserve some plants. Placing plants in a plastic bag until an alternative can be worked out should keep them alive and free of any of the contaminants that may be found in soil.

Sheets of polyethylene can be used to form a miniature hydroponic system outdoors. The polyethylene is cut into strips about 15 inches wide and of whatever length is required. The plastic is rolled to form a tube and laid down on any plot of ground having a gentle slope. Along the length of each tube, at regular intervals, holes are cut and plants gently inserted. Nutrient solution is pumped down through each roll, flowing over the roots of the plants and collecting in a reservoir excavated at the end of the row. This method, known as the *layflat,* is generally implemented as a series of rows, arranged parallel to each other on a slope. A pump located at the top of the slope is used to send nutrient solution through connecting passages to each of the rows. Another pump is used to pull nutrient solution back up from the reservoir, so that it can be recycled. This method requires an ample piece of ground to be economically worthwhile and can really only work when implemented on a large scale.

Aeroponics is one of the more unusual variations of the hydroponic principle. Plants are grown in tubes or containers arranged either vertically or horizontally, through which a spray of air and nutrient solution is continuously pumped. The plants are inserted into the containers or tubes so that their roots are inside, but their stems, leaves or flowers are all projecting above the surface of the unit. The roots are constantly moistened by the solution and there is always an adequate supply of oxygen. Of course, a good deal of machinery is required to draw in sufficient amounts of air, mix it with the solution, and release it with enough force for it to penetrate the entire unit.

The *hanging basket method* makes use of baskets having permeable bottoms made of netting or of some porous material which is sufficiently sturdy to contain aggregate material and plants. These baskets are suspended on heavy wires and are rigged so that they can be lowered into containers of nutrient solution. After having been allowed to absorb the solution, the baskets are again hoisted up. Since they are positioned above the containers, any solution that flows out of the units will simply fall back into the reservoir.

THE BENGAL SYSTEM

Only a few individuals can be said to have a comprehensive understanding of the hydroponic field. James Sholto Douglas must certainly stand foremost in their ranks. The inventor of the Bengal system of hydroponics, he is also the author of three books on the field, including the most complete survey of the entire science of hydroponics. While Douglas's system was designed specifically for conditions prevailing in the Bengal region of India, the requirements that shaped it have produced a system perfectly adaptable to use in the home.

Douglas had to develop a hydroponic system that was inexpensive to build, easy to assemble, simple to run, and functional in locations where ordinary methods of gardening were impractical or impossible. Of course, though the reasons for having such a system differ greatly, such a system is as attractive to us as it is to the Indian farmers for whom it was developed. Most of us, whether living in a house or an apartment, must deal with having a very limited space in which to garden, cannot afford to experiment with complex or expensive systems and have little technical expertise to handle demanding horticultural projects. Simplicity benefits us all.

Almost any sort of container (except one made of galvanized iron) can be used to set up a small "hydroponicum," or hydroponic unit. A container as small as a flowerpot can be pressed into service. Any kind of aggregate is acceptable, so long as it is capable of retaining moisture without becoming so compacted as to suffocate a plant's roots. As with all other hydroponic systems, a reliable method of supplying the container with a water and nutrient solution is necessary.

The Bengal system differs from others primarily in its very adaptable attitude towards containers and materials. Mr. Douglas has produced successful units from such materials as clay, stones and bamboo. He describes the first successful Indian hydroponic unit, constructed in 1946, as having "...fairly small wooden containers made by sawing old packing cases in two; one large sector consisting of beds with non-erodible mud plaster walls on a concrete base; and finally other tanks made entirely from mud plaster. Later, a considerable number of ordinary plant pots and a few metal containers were brought into use.... Irrigation was supplied both by a bamboo piping system and from rubber hose" (*Hydroponics, The Bengal Method*, pp. 79–80).

The aggregate was composed of a mix of five parts of rock chippings and one part of rock dust. Large mountain boulders, which are plentiful in the nearby hills, were broken up with hammers to the required size. The action of breaking these rocks up yielded both chips and rock dust, and these two materials were then mixed together and poured into the containers to a depth of eight inches.

Since that first successful experiment, hydroponics has proved to be of great use to Indian farmers. Large units now yield considerable amounts of food crops. The techniques involved are much more sophisticated, but they are also easily duplicated.

The most important facts to be learned from the success of the Bengal system are that the principles of hydroponics can be adapted to a wide range of conditions and that a functional, effective system can be built from inexpensive materials, often those closest at hand. As Douglas notes, there are no secrets to successful hydroponic gardening. All that is required is "a careful study of the particular place, plus ingenuity, improvisation and sensible adaptation" (p. 29).

So many developments in agriculture require heavy investments in equipment and trained personnel to operate them. Hydroponics needs neither, and is thus of major importance to developing nations short on cash and technical expertise. The experiments in India provide proof that hydroponics is a workable alternative method of agriculture not just for industrial nations but for the entire world community.

THE MITTLEIDER METHOD

The Mittleider method combines the principles of hydroponic culture and soil gardening to produce a unique, innovative method of raising food crops. It provides further proof of the adaptability and effectiveness of hydroponic principles and offers an intriguing alternative to conventional farming methods. According to the literature on the subject—including two books by its inventor, Jacob R. Mittleider—the system has been tried in a variety of climatic conditions throughout the world, and has proved itself capable of outproducing other gardening methods. The technique also lends itself to situations in which conventional gardening practices are impractical or impossible. In addition, it is an inexpensive system, requiring little machinery since most of the work is done by hand. It also makes such good use of water and nutrients that supplies of both can be reduced below the levels required in conventional agriculture.

According to Jacob Mittleider, the system is "based on maximum utilization of space, time and resources. Crops are large because plants are grown close together, nourished by supplemental feedings of mineral nutrients, as in hydroponics, but with no special equipment" (*More Food from Your Garden,* p. 29). The Mittleider method differs from hydroponics in that plants are grown in a specially prepared soil mix in the ground so that the crops are given access "to the natural soil for nutrients as yet unknown or that, while not essential to plant growth, are useful in human nutrition" (p. 29).

The focus of the method is the "grow box," a bottomless container with wooden or cement walls, built firmly into the ground. A grow box can be of any size or shape, to adapt to available space. The most frequently employed box is thirty feet long and five feet wide, with walls eight inches in height. These boxes are then filled with a special soil mix compounded of sawdust and sand, perlite and peat moss, or fine sand and peat moss. According to Mittleider, this mixture offers the advantages of excellent drainage and good aeration. In short, it serves the same purpose (and uses some of the same materials) as the aggregate in hydroponic culture. Because the ingredients of the soil mix contain few nutrients, these are supplied by the gardener. Again, we're talking about a process similar to hydroponics. However, at this point the method begins to part company with hydroponic techniques. The roots of the crops are allowed to press downwards through the mix and into the soil. This, as Mittleider notes, allows the plants to "absorb many extra important minerals."

Grow boxes can be easily worked with a few simple tools. The design allows you access to every plant. It also frees you from having to walk about the rows since you can simply lean over to reach any plant. While not a serious problem, walking on soil can cause the ground to become hard and your shoes can spread the seeds of various weeds or the spores of diseases among your plants.

How much can a grow box produce? According to Mittleider, "four grow boxes can provide a nearly continuous harvest of food much of the year for a family of four. . . . Ten grow boxes can provide for a family's needs and a large surplus for selling or sharing. Twenty-five to fifty grow boxes can provide for complete economic sufficiency" (p. 35).

After the soil mix has been prepared and poured into a grow box, nutrients must be thoroughly mixed into the soil. Thereafter, additional applications of fertilizer are accomplished by spreading dry, granular nutrients over the surface of the aggregate. A hose is then turned upon the unit and the water serves both to dissolve the granules and carry the nutrients down to the roots of the plants. Nutrients must be applied to the unit once each week. These nutrients, which differ from those required in the initial application, can either be purchased separately and mixed by the gardener or they can be obtained in premixed preparations.

The Mittleider method has been designed quite specifically for food crops. Any vegetable can be raised in a grow box, but you might want to steer clear of perennial vegetables for they must be allowed to remain in place year after year, taking up space that could be better utilized to raise other, fast-grown crops, crops that could be picked in time to replant the space with yet more plants.

This concept of containerized gardens has, I believe, wider applications than just for food production. I don't see why the concept of the grow box cannot be adapted to the growth and display of flowers, shrubs, herbs—in short, anything green you desire.

The use of grow boxes filled with aggregate would greatly simplify what is frequently the hardest, least-appealing gardening task—preparing the soil in a garden, making it sufficiently loose and rich to insure proper plant growth. Using an aggregate in place of soil would greatly simplify the cultivation of flowering plants, shrubs, or rose bushes. As with vegetables grown by the Mittleider method, very exact amounts of fertilizer could be applied to the aggregate, with the assurance that a high percentage would be absorbed by the plants instead of being lost in the soil.

You could also create some unique, attractive displays within a grow box, using the units as the focus of the garden. Because I have a special fondness for the plant, I have been considering building a circular grow box and filling it with tulips. The sides of the unit could be disguised by growing herbs as a border, or by creating a bank of earth all around the unit.

Only the kind of soil in which plants are grown and the method of applying nutrients are altered by the Mittleider method. You must still observe the needs of the plants for adequate light, for humidity, and for an acceptable temperature range. Should you wish to do so, you could gain greater control over the supply of warmth and humidity by erecting greenhouses over the grow boxes. These greenhouses need not be elaborate and they should not be expensive—because the food crops will need all the light they can get, you won't be able to place benches about them. You cannot fully utilize the space within the greenhouse, so don't overextend your budget in building one. Jacob Mittleider includes plans for inexpensive greenhouses assembled from polyvinyl chloride piping and sheets of plastic in his book. The Bibliography lists more books explaining the Mittleider method, and several others devoted to greenhouse construction.

You could also buy a prefabricated greenhouse, but one large enough to cover a thirty-foot-long grow box would be very, very expensive. You can circumvent this problem somewhat by building several smaller grow boxes, by rearranging the shape of the boxes to fit into a greenhouse. The simplest method would be to create arches of polyvinyl chloride piping two feet tall and five feet wide. Arrange the arches in a line over the grow boxes with the arches straddling the unit. Stretch plastic over the arches. What you will have is a miniaturized version of the greenhouse. Anchor the plastic with soil or bricks, and simply lift back each portion of the plastic sheets as you need to reach the plants below. Of course, you won't be able to raise any vegetables requiring tall supports, but you will be able to grow any other crop. The only potential problem would be a lack of adequate ventilation, and you could remedy this by removing the plastic entirely during the hottest hours of the day.

The Mittleider method somewhat simplifies the process of growing vegetables hydroponically. Because so little equipment is required, it also costs somewhat less than it would to raise vegetables in a full-fledged hydroponic unit. Since the plants can be placed more closely together than if they were being grown in the soil (because the plants are receiving all of the nutrients and water they need, their roots need not spread out and contest with each other's root systems), a higher yield can be gained than from plants raised in soil. The method, composed of a variety of aspects of other techniques, is a viable alternative, and in some cases may be preferable to either growing crops in soil or in a hydroponic unit.

PORTFOLIO OF PLANS FOR SMALL, HOME HYDROPONIC UNITS

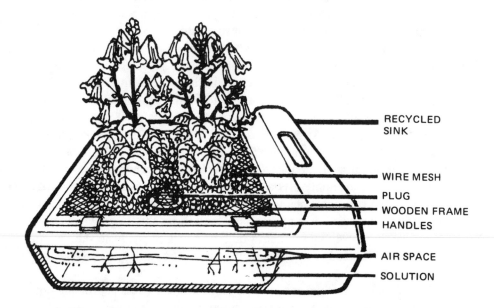

RECYCLED SINK

WIRE MESH

PLUG

WOODEN FRAME

HANDLES

AIR SPACE

SOLUTION

An example of what happens when you let your imagination go. This sink makes a perfectly acceptable hydroponic unit, and very little construction is necessary. Over the wire mesh a thin layer of wood chips, shavings, or some other aggregate is scattered. In this case, the plants are actually growing in solution, not in an aggregate. However, there's no reason why the sink couldn't equally well be filled with an aggregate such as vermiculite or gravel, and the wire mesh arrangement done away with.

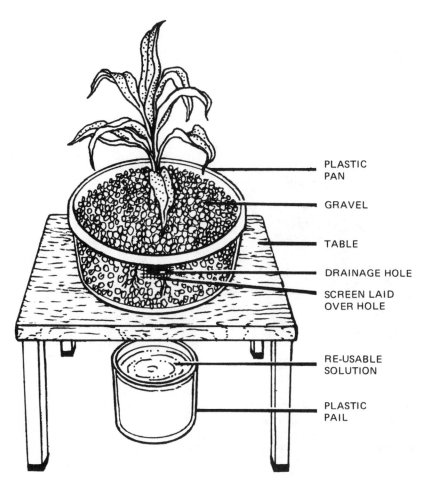

PLASTIC
PAN

GRAVEL

TABLE

DRAINAGE HOLE

SCREEN LAID
OVER HOLE

RE-USABLE
SOLUTION

PLASTIC
PAIL

Another very simple, very inexpensive idea for a unit. You can knock together a simple table from pieces of scrap wood—it need only be high enough for a plastic pail or bucket to sit underneath. The container (it can be of any sort) in which you place the plants should be deep enough to allow a depth of six to eight inches of aggregate. Pour in sufficient solution to come halfway up to the level of the aggregate. Allow the solution to pour through the drain you've fixed in your table and into the pail beneath. Pour the solution from the pail back into the unit two to three times each day.

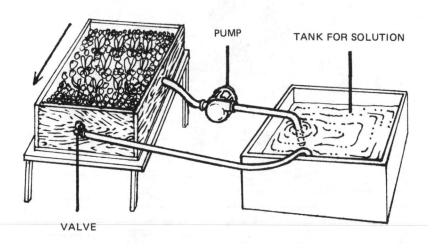

PUMP

TANK FOR SOLUTION

VALVE

A more complex unit, using a tank to carry solution into the container. Notice that the container with the plants is arranged so that the solution will slope downwards towards the hose carrying the solution back into the solution tank. Shards of crockery or half-tiles can be arranged under the growing medium to insure that the solution flows freely through the medium and back towards the hose.

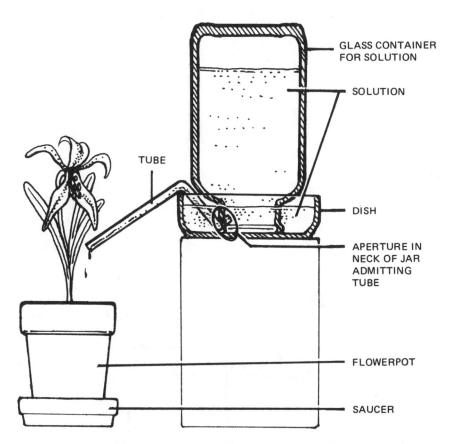

GLASS CONTAINER
FOR SOLUTION

SOLUTION

TUBE

DISH

APERTURE IN
NECK OF JAR
ADMITTING
TUBE

FLOWERPOT

SAUCER

An automatic unit can be constructed without the use of much complex machinery. In this case, a piece of capillary tubing, affixed to the neck of a jar and arranged at the right angle will supply a constant, unvarying amount of solution to a unit. This idea makes an interesting experiment, but is impractical for large-scale cultivation. A groove must be cut in the neck of the jar you are using to allow the tubing to be slipped in. The tube should penetrate into the jar to a length of two inches. It should be bent downwards at the lip of the saucer. The steeper the angle of the tube, the faster the solution will flow. Getting the unit to work properly will require a good bit of fiddling.

Fill the jar with solution, put the tube in place, place a saucer or dish over the neck of the jar, and quickly invert the jar so that it is standing upsidedown in the dish. You may have to suck the tube to get the solution flowing. After it has started, it should continue dripping steadily until the supply of solution is exhausted. If the growing medium is becoming waterlogged, you will have to adjust the angle of the tubing to slow down the rate at which the solution is dripping out. Or you could alter the angle of the tubing, turning it upwards after a set amount of time has passed, to shut off the flow of solution. You could repeat this procedure several times a day, turning the tubing down each time for a long enough period to allow the medium to be thoroughly moistened.

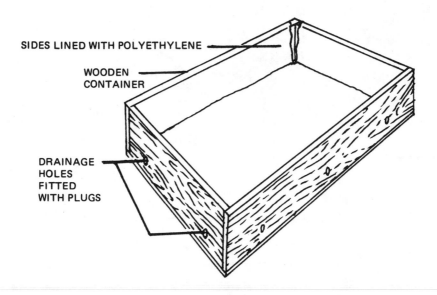

SIDES LINED WITH POLYETHYLENE

WOODEN
CONTAINER

DRAINAGE
HOLES
FITTED
WITH PLUGS

The unit illustrated here is a simple wooden box, of any acceptable length, having a depth of from six to eight inches. The entire interior of the unit must be lined with polyethylene sheeting or any other waterproof material. One wall of the unit should have two small holes drilled through. The holes should be located half an inch from the bottom of the box. The holes must be fitted with removable plugs.

After the unit has been outfitted with its waterproof lining and drainage holes, pour in large pieces of clean gravel or shards of crockery (from broken flowerpots) to a depth of one inch. In this way you can make certain that the drainage holes will not become clogged. Also, this coarse bottom layer provides additional aeration for the plants.

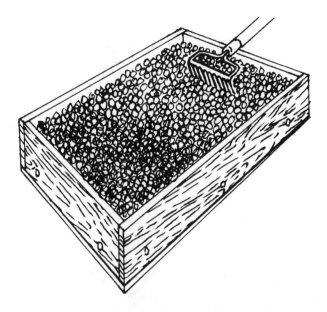

Once you've done this, pour in the growing medium you've chosen, filling the unit in until the material reaches to a point one-half inch from the top of the container. Use a clean rake to even the material out. Then use a hose or a watering can to apply sufficient water to thoroughly moisten the medium. This encourages the medium to settle. A half-hour after the watering, remove the plugs and allow any excess water to drain out. You'll need one or several receptacles ready to catch this drain-off. This initial watering also allows you to test your drainage system. If no moisture emerges, apply more water. If no water is forthcoming from the drainage holes, there

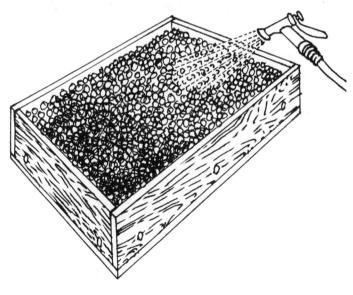

is a problem somewhere in your system. Perhaps the holes aren't big enough. Perhaps your waterproof lining around the holes has not been pierced. Or perhaps you have insufficient drainage material in the bottom of the unit. A block of wood placed under the end of the unit opposite the drainage holes will facilitate proper drainage.

PIPE OR GARDEN HOSE

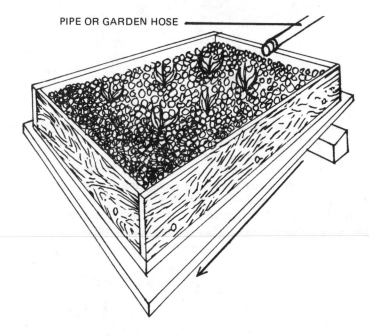

After the unit has been filled and the medium moistened, you're ready to sow seeds in the unit or fill it with seedlings or mature plants.

3. Home Hydroponics

Once you have settled on the kind of unit best suited to your needs and interests, you'll have to decide whether you want to build it yourself or buy a kit. The decision depends largely on the time and money available to you. After you have your unit, the most important chore you'll face in running it is the proper preparation and application of the nutrient solution. However, once you have established a successful pattern in this matter, there will be little reason to deviate from it.

After your pattern of tending to the unit has been worked out, you can turn to the equally important matter of tending to the plants in the unit. This is largely composed of supplying sufficient warmth, light and moisture for the plants.

This chapter, then, is concerned with the specifics of your twin responsibilities when raising plants hydroponically: the proper maintenance of the unit and the proper care of the plants.

ON BUYING HYDROPONIC COMPONENTS AND KITS

When I began reviewing the literature issued by manufacturers and suppliers of hydroponic components and complete hydroponic kits, I was amazed to discover how many companies were working in the field and how much they had to offer. You can find, with a bit of looking, everything from simple $15 containers to $20,000 greenhouses that seem to offer the last word in automation. There are a wealth of kits and parts, a cornucopia of nutrient solutions. This embarrassment of riches is both pleasing and confusing. While you have a great range of products to choose from, it's not easy to make a choice. Fortunately, some of the kits and accessories are beginning to be sold by garden supply centers and plant shops, so you can see at least some of them firsthand. Many of the manufacturers have also made available detailed descriptions of what they have to offer.

While all this may help you make your choice, your decision on what to buy is still finally your own. Of course, the first thing you must do is to decide how large a hydroponic unit you need and have room for. How much you can afford to spend is also an important consideration. (A $20,000 unit might be great fun, but it is beyond both the needs and finances of all but commercial gardeners or fanatics.) Once you've settled these two essential questions, the list of possibilities will be greatly reduced.

You can handle your purchases in one of two ways. You can build your own unit, restricting your purchases to those things you cannot make or otherwise adapt from materials at hand (such as a pump). Or you can buy an entire hydroponic unit, including a container, an aggregate, and in many cases whatever sort of pumping device is necessary. Many kits also include an initial supply of nutrients. While the small beginner kits are generally fairly simple, larger kits of far greater sophistication are also marketed, along with a number of accessories.

If you're planning on starting out with just one unit in your home, you can as easily buy as build. However, when you build a unit you can adapt it to the particular circumstances of your home—you can more closely fit the unit to your needs. If you intend to have several large units in your home, you would do better to build them, as it would be less expensive to purchase large quantities of materials than to buy manufactured containers. Aside from these considerations, the decision is largely one of convenience. If you're uncomfortable working with your hands, go shopping for a unit. If you prefer to build things yourself, by all means do so.

A list of manufacturers and suppliers of hydroponic products and the items they offer can be found in the Resources section of this book. Most offer catalogs or descriptive brochures. When you're shopping for hydroponic products, the same rules of careful purchase apply as for any other item. Read the literature thoroughly. Compare guarantees. Compare prices and materials. If at all possible, try to inspect some of the product firsthand. Remember that no matter how good a product is, it's the businessman's job to sell it. It's up to you to receive his claims cautiously and examine them carefully. Before you buy anything, read up on the subject. Develop a rough idea of the size of the unit you want and what you want to grow in it. Much of the promotional material won't make sense unless you understand the basics of hydroponics. So read. Find out about a number of units. Compare them carefully. Then and only then should you spend your money.

THE NUTRIENT SOLUTION

No matter what sort of hydroponic unit you've decided to set up, the plants in it will have to be regularly supplied with nutrients. As already noted, when you begin gardening hydroponically you assume the responsibility of supplying the nutrients plants normally draw from the soil. Because the supply of nutrients in a given patch of soil can fluctuate wildly, and because the process of applying nutrients in the form of compost and fertilizers can prove very inexact, hydroponics allows you to make nutrients available to plants in a much more precise manner.

The usual method for applying nutrients to a hydroponic unit is to dissolve individual nutrient "salts," or a commercially mixed nutrient preparation, in water. This solution is then pumped into or poured over the growing medium.

There are a number of premixed nutrient mixes now on the market. The only preparation required is to measure out the appropriate amount and stir it into water. If you have just begun working with a hydroponic unit, or if your hydroponic gardening is restricted to a small unit, it will probably be easier for you to use one of the commercial mixes. However, there is nothing difficult, nothing esoteric, about mixing your own preparation from the individual nutrients required. They are all available at garden supply centers.

What does each of the nutrients do?

NUTRIENT	FUNCTION
Nitrogen	Influential in the production of leaves and in the growth of the stem. An essential ingredient in protoplasm—the basic stuff of life found in each living cell—and thus quite literally an indispensable part of a plant's composition.

NUTRIENT	FUNCTION
Phosphorus	Needed in the development of flowers and fruits. Also encourages growth of healthy roots.
Potassium	Used by the cells of a plant during assimilation of energy produced by photosynthesis.
Calcium	Spurs root growth. Also facilitates a plant's absorption of potassium.
Magnesium	A component of chlorophyll. Also active in the process of distributing phosphorus throughout a plant.
Sulphur	Joins with phosphorus to heighten the effectiveness of that element. Also used in the production of energy.
Iron	Important in the production of chlorophyll within a plant.
Manganese	Aids a plant in the absorption of nitrogen.
Zinc	Necessary component of the energy transference process in a plant.
Boron	While it has been established that boron is needed in minute amounts, it is not known precisely *how* boron is used.
Copper	Needed in the production of chlorophyll.

Sources of nutrient salts

The elements required by plants are available in a number of commercial preparations. Any of these preparations are acceptable, but since the amount of the element present will vary from preparation to preparation, so must the amount you will use in your mix.

ELEMENT	SOURCE
Nitrogen	*Potassium nitrate.* Source of both potassium and nitrogen. Very soluble, readily available, and keeps well. *Sodium nitrate.* Source only of nitrogen, since sodium is not required by plants. Inexpensive, very soluble, stores well if kept in a tightly lidded container in a dry location. *Calcium nitrate.* Contains both calcium and nitrogen. Less easy to store. Should be used only when other preparations are unavailable.
Potassium	*Potassium sulfate.* Very soluble, stores well. Should be your first choice. *Potassium chloride.* Can be used when potassium sulfate is not available, but can prove harmful if used for more than a few days since the chlorine in the mix makes it potentially harmful to plants.
Phosphorus	*Treble superphosphate.* The best choice, although plain super-phosphate is also acceptable. Treble superphosphate also supplies calcium. Both sources may contain impurities of trace elements. Superphosphate accidentally splashed on leaves will cause harmless white blotches to appear.
Magnesium	*Magnesium sulphate* (Epsom salts). Inexpensive, dissolves readily in water, and stores well. The only fertilizer salt to be found in medicine cabinets and on the shelves of drug stores. *Magnesium nitrate.* May also be used, but is more expensive. Since Epsom salts are so cheap and so readily available, there is little reason for the home gardener to bother with other sources.
Calcium	*Calcium sulphate.* Both gypsum and plaster of Paris are composed of calcium sulphate. The plaster of Paris dissolves more readily.
Iron	*Ferrous sulphate, ferric chloride* or *ferric citrate.* All prove acceptable as sources of iron. Ferrous sulphate and ferric chloride will dissolve in cold water, while ferric citrate will only dissolve in hot water. Ferric citrate will remain in solution longer than the others and is also more stable at high pH than the others. It is often preferred for these reasons.
Manganese	*Manganese sulphate.* The most commonly used source of manganese. Must be kept in a dry, tightly lidded container. *Manganese chloride.* A less common salt, can also be used.
Boron	*Boric acid.* The best source of boron. *Borax.* May also be used in an emergency. *Copper sulphate* and *zinc sulphate* both contain supplies of boron. Boron and manganese are often found in sufficient quantities as impurities in other nutrient salts.

Buying and storing fertilizer salts

Garden supply centers generally carry a selection of fertilizer salts. The bags in which the salts are packaged will carry figures denoting the amounts and identities of the elements in the product. The order of elements in such a list is unchanging: nitrogen comes first, then phosphorus, and finally potassium. The elements are sometimes represented in both abbreviations (N for nitrogen, P for phosphorus, K for potassium) and numbers, though usually only numbers represent the percentages of the elements in the product. A label reading 10–0–0 would then indicate that the preparation was composed of nitrogen equalling ten percent of the material, with the remaining ninety percent of the preparation being composed of inert materials or trace elements. A reading of 10–8–0 would indicate that in addition to ten percent of the product being composed of nitrogen, eight percent of the remaining material would be phosphorus and the remaining eighty-two percent would be inert. Many packages carry more detailed annotations covering each of the trace elements present in the preparation.

Fertilizer salts are manufactured by a number of companies and are available in a variety of sizes. Buying very large bags of each of the necessary salts is certainly more economical, but it may be awkward for you to transport or store bags of ten, twenty-five or fifty pounds each. However, if you have a fairly large hydroponic setup, such as a greenhouse, it is more logical and definitely less expensive to purchase in quantity.

Whatever the size of your unit, it is a good idea to transfer the materials to containers having tightly fitting lids. Label the containers so that no potentially damaging mistakes can be made and keep them in a spot receiving little traffic. If there is the chance of paws or small hands knocking the lids off and dipping into the salts, take steps to isolate the containers or to make their lids entirely secure. I know of one gardener who salvaged large drums from a nearby warehouse, cleaned them out and marked each with a broad band of tape. Different colors of tape were used to denote different salts. The drums then had their lids firmly fixed in place and were stored in a small locked utility room. It is essential, whether you store the salts in containers or leave them in their original bags, that they be kept clear of *any* moisture.

Mixing it yourself

There are scores of formulas for nutrient solutions. Some of them are quite complex and use a great many ingredients. The formula given here is well suited for beginners since it contains only a few readily available salts.

FERTILIZER SALTS	QUANTITY (IN OUNCES)	NUTRIENTS SUPPLIED
Ammonium sulphate	1½ (43 grams)	Nitrogen, sulphur
Potassium nitrate	9 (255 grams)	Nitrogen, potassium
Monocalcium phosphate	4 (113 grams)	Phosphorus, calcium
Magnesium sulphate	6 (170 grams)	Magnesium, sulphur
Calcium sulphate	7 (198 grams)	Calcium, sulphur
Iron sulphate	Enough to cover the nail of your index finger	Iron

Use a large, clean bowl or other container to mix the ingredients. Weigh out each of the materials carefully, then pour them into the bowl. After all of the salts have been added, use a chemist's pestle, a broad wooden spoon, or any blunt, clean instrument to break up any crystals in the salts. Because there is such a miniscule amount of iron sulphate, it is preferable to crush it before adding it to the mix. After it has been added, stir the materials thoroughly, breaking up any additional crystals you find. In this way the iron sulphate will be distributed throughout the mix.

When you are finished, you should have a rather fine powder. Store the mix in a clean, dry container with a lid. It is essential that the formula be kept perfectly dry until it is to be dissolved in water and applied to the plants. Indeed, even the components must be kept dry before they are combined in a mix. Keep any extra supply of the various ingredients in tightly sealed containers until they are to be used.

You will need to use only a small amount of this formula each time—one third of an ounce (10 grams) to each gallon of water. In practical terms, one level teaspoonful of formula stirred into a gallon of water will prove to be sufficiently exact. Be sure to stir the water until the powder has completely dissolved.

The fertilizer salts listed above are not the only ones that you can use. For instance, another formula calls for the use of:

FERTILIZER SALTS	QUANTITY (IN OUNCES)	NUTRIENTS SUPPLIED
Sodium nitrate	12½ (355 grams)	Nitrogen
Potassium sulphate	4 (113 grams)	Potassium, sulphur

FERTILIZER SALTS	QUANTITY (IN OUNCES)	NUTRIENTS SUPPLIED
Superphosphate	5 (142 grams)	Phosphorus, calcium
Magnesium sulphate	3½ (100 grams)	Magnesium, sulphur
Iron sulphate	Enough to cover the nail of your index finger	Iron

The same elements are supplied, but the salts from which they are released differ in composition from those given in the previous formula. Because the salts are different, so are the quantities in which they are added to the mix. However, the outcome is the same and the same amount must be added to water as is called for in the first formula—one-third of an ounce or (for the less exacting gardeners) one level teaspoonful to each gallon of water.

These formulas provide the major elements required by the plants. There are also, however, trace elements which are needed in much smaller amounts, but needed nevertheless. Fortunately, these elements are frequently present as impurities in both the various fertilizer salts listed above and in your water supply. If you remain concerned about their possible absence, or if your plants show definite signs of a deficiency of one of these elements, you can prepare the following formula to be added to the nutrient solution.

FERTILIZER SALTS	QUANTITY	INSTRUCTIONS FOR USE
Manganese sulphate	1 level teaspoon	Mix these ingredients together, and store the mix in a dry container having a lid. When you require some, dissolve ½ teaspoon of the mix into 1 quart of water. Use 1 liquid ounce of the solution for every 3 gallons of water applied to the plants.
Boric acid powder	1 level teaspoon	
Zinc sulphate	½ level teaspoon	
Copper sulphate	½ level teaspoon	

After you have removed an ounce of the trace mix from the quart you have prepared, discard the remainder of the quart. The trace mix cannot be kept in a liquid for any length of time and retain its potency. It can, however, be used the same day as it is prepared. Instead of simply pouring any unneeded solution down the drain, take it outside and distribute it among your other plants. Spread it around, taking care not to pour too much on any one spot.

These formulas should prove adequate for use in home hydroponic gardens. In the most sophisticated hydroponic installations, computers are used to design very exact formulas for each kind of crop being grown. These are the result of tests in which every conceivable problem and variation has been examined. Even the water to be used is subject to analysis. Every facet of such operations is almost constantly monitored, and additions or deletions to the formulas are regularly carried out. Such projects have produced large, healthy crops and a mass of useful information.

Watering practices

How should the nutrient and water mix be applied?

The two basic systems of applying a water and nutrient mix to a hydroponic unit are active and passive. In an active system, the solution is poured in and then, by means of gravity, allowed to drain off. This way you can collect the nutrient solution and reuse it. Active systems require growing mediums that drain evenly and rapidly, such as smooth gravel.

In a passive system, a group of wicks anchored in the medium and stretching down to a water reservoir draw moisture upward into the growing medium to reach the roots of the plants. Passive systems require materials that retain moisture longer, such as a mix of vermiculite and sand.

Active systems have more moving parts, such as pumps. Passive systems are somewhat more complex to set up but are less likely to malfunction. However, my experience shows that passive systems are slightly more vulnerable to pest and disease problems. I believe the most effective systems for home use are active systems. Large-scale commercial growers, able to draw on a number of heavily automated components and on a good deal of human labor, have had great success with passive systems in large greenhouses.

What should the water temperature be?

The water you use in your unit should never be colder than 60°F., nor warmer than 80°F.

How frequently should nutrient solution be applied to a unit?

Often enough so that the growing medium never dries out, but not so frequently that it becomes waterlogged. The only way to establish the appropriate frequency is to learn to "read" the medium. How long does it take to dry out? Once you know that, you will know how frequently to apply solution or when to set a timer so that solution is automatically supplied at the right time.

How much solution should be applied?

Enough to insure that the aggregate or growing medium is always moist. James Sholto Douglas writes that the medium should always be as moist "as a damp sponge that has been squeezed out lightly" *(Beginner's Guide to Hydroponics,* p. 53). A unit can be said to be waterlogged when puddles are visible on the surface of the growing medium, and the medium has become soggy or collects in wet lumps around the plants. Again, the only way to find out how much solution your unit needs is to observe it. As you gain experience, you will learn how much solution to apply. This experience will come quickly. In the meantime, try to neither flood nor parch your unit.

Flushing the unit

No matter whether you are raising plants hydroponically in a flowerpot, in a large unit, or in trenches in a greenhouse, it is a good idea to flush the container out every two weeks by pouring in plain water. This will serve to remove any accumulations of unused nutrients in the growing medium. Allow all of this water to drain out, then discard it.

Re-using solution

Can you re-use solution?

Yes, but not for an extended period of time. While very large hydroponic units such as those run at experimental stations may re-use solution for as much as a week, they have the equipment and the skill to determine if any of the nutrients have been depleted.

If you choose to re-use solution (and it is an economical and ecologically sound choice) I think you should limit the use of any one batch to three or at the most four days. If your plants begin to show signs of nutritional deficiencies, begin substituting fresh batches of solution more frequently, say every two days. As with watering practices in general, you are the best judge of what's needed.

CONCERNING pH

The term pH is much used and frequently misunderstood. A pH reading refers to the acidity or alkalinity of any given material that is wet or water soluble. Hydroponic gardeners are primarily concerned with securing pH readings for the water used in mixing up a batch of solution, or for the finished solution itself.

There is a scale for pH readings, running from 0 to 14. A reading of 7, commonly associated with pure water, indicates that the numbers of acid ions are equal to the number of alkaline ions present. Any reading above 7 indicates that the alkaline ions are dominant, and any reading below 7 means that the acid ions are present in a greater number. Each number above or below 7 indicates a *tenfold* increase in either the acidity or alkalinity. A pH reading of 4 would be *ten times* as acid as a reading of pH 5. If a solution is only one figure away from the pH reading recommended for your solution, that solution is ten times more acidic or alkaline than it should be.

Why does this matter? If the solution is either too acidic or too alkaline, certain of the vital nutrients in the nutrient solution will be precipitated into insoluble salts, a form in which plants cannot absorb them. And without these salts the plants will begin to exhibit symptoms of various deficiencies. For example, a pH level far below 6 would cause a deficiency in calcium to occur. Injuries connected to such a deficiency include damage to the root system, burning of tissues in the tip ends of roots, wilting of foliage, and the development of patches of rotten tissue. A solution that is too alkaline will interfere with the plant's absorption of iron, causing the onset of chlorosis. Other difficulties may also occur. But having said this, we should also note that plants are rather adaptable in this matter, as in many others.

What should the pH be? A reading of from pH 6 to pH 6.5 is generally recommended, but plants can adapt to levels that read as low as pH5 or as high as pH 7. A slightly acidic solution is preferable. In most cases, the water you're using or the solution you mix will come up with a reading close to the norm. Problems usually occur only after a batch of solution has been used for several days. As the nutrients are absorbed from the solution by the plants, the remaining solution will become increasingly alkaline. Fortunately, it is easy to take a pH reading and correct any imbalance.

The solution used in commercial hydroponic units and large-scale experimental stations is rigorously monitored and regularly analyzed by computer. You need not be so dutiful. Indeed, some gardeners tending hydroponic units run pH checks only once every several weeks or when they suspect the plants are not as healthy as they might be.

The pH of the water you intend to use for the unit should be determined before any of the nutrient salts are mixed into it. Nitrazine paper, available at pharmacies and drug stores, can be used to carry out a quick but sufficiently accurate test of the pH. The test kits sold to owners of swimming pools can also be used to identify the pH of your water. Or, if you desire to carry out very accurate tests, pH test kits designed for gardeners are manufactured by several firms (see the Resources section for a list).

Some of these kits include information on how to raise or lower an unacceptable pH level. Among the most easily obtained ingredients which can be used to lower a high pH is aspirin. Two aspirin tablets per gallon of water will lower a pH reading of 8 to something close to pH 6. A teaspoon of distilled white vinegar added to each gallon of water will accomplish the same purpose. It is much more common to find water with a high pH than with a pH range falling below 5.5.

A low pH can be corrected by adding either potassium hydroxide or sodium hydroxide to the water. Only very small amounts of either should be needed.

Because the hydroxides are caustic, they should be used only with great care. *Never* touch hydroxides when your hands are wet. Indeed, you should try and handle them as little as possible. Keep any reserves of hydroxide in a container having a tight lid and store the container in a spot where it is unlikely to be accidentally knocked over.

A PRIMER OF PLANT CARE

Raising plants hydroponically alters the way in which they receive necessary supplies of water and nutrients, but it does not alter their need for these things. Nor does it affect a plant's need for sufficient light, warmth, and humidity. Whether plants are grown in soil, sand, gravel, or some other medium, they need light and heat just as much as they need water and nutrients. However, different kinds of plants need differing amounts of these elements. If they cannot receive sufficient amounts of any of these, or if they receive too much, the plants will surely sicken. If the conditions are not corrected, they will die. Before you buy any plants, before you set up your unit, you should determine what amounts of light, warmth, and humidity are available in your rooms. After you have established these figures (and you can do so quickly, accurately, and inexpensively) you can either restrict your plant purchases to those species that will adjust to such conditions, or you can set about altering those conditions to broaden the variety of plants you can raise.

The first thing to do is to establish what these conditions are.

Warmth

The vital process of photosynthesis can only take place when the atmosphere around a plant is sufficiently warm. Should the temperature fall too low, or should it fluctuate wildly, photosynthesis may occur infrequently, incompletely, or not at all.

The tropical conditions in which many plants evolved had a profound influence on their growth mechanisms. Plants developed in a way that made them dependent on a supply of moisture in the air and on temperatures that were usually rather high. The benign influence of the tropics made a diverse plant life possible, but also shaped it to its conditions. Plants have succeeded in colonizing other, more demanding environments only by becoming hardier, by making certain adaptations to carry them through periods of low temperatures. But plants have never lost their dependence on warmth—even if that warmth is only seasonal.

If the area around your plants is too cool they may grow more slowly or cease growing altogether. Use an inexpensive garden thermometer to take a series of readings in your rooms, checking each location in which you have considered setting up your hydroponic unit. Take a reading in the morning, another in the afternoon, and yet another at night. Most plants will do very well in a spot having a daytime temperature no lower than 60 to 65 °F., although a preferable range would be from 70 to 75 °F. Night temperatures as low as 50 °F. are perfectly acceptable. Indeed, plants must have a night temperature at least ten degrees lower than the daytime temperature. A high temperature maintained through the night would cause the plants to get signals mixed and go on attempting to produce energy when they should be assimilating all of the energy produced during the day.

Reading the relative humidity

The term "humidity" is used to refer to water suspended in the air as vapor. "Relative humidity" refers to the percentage of moisture in the air as it is measured against the total amount of moisture which can be sustained in the air without being precipitated into rain or fog. A reading of 40 percent relative humidity would mean that the air sampled contained 40 percent of the moisture it could sustain before the moisture would grow too heavy, condense and be precipitated.

Warm air can carry a greater amount of vapor without the moisture being precipitated. A reading of 40 percent relative humidity at an air temperature of 72 °F. (which is, incidentally, an excellent combination of humidity and warmth for plants) would indicate the presence of *more* water vapor than a reading of 62 °F. and 40 percent humidity. The term "relative" is thus quite expressive. The amount of humidity present really is relative to the temperature reading.

Humidity is important to plants because so many species evolved in tropical or semitropical areas where the air is always heavy with moisture. Even plants that have evolved or, over the course of thousands of years, adjusted to conditions of lower humidity and temperatures experienced higher levels of humidity in their natural settings than can be found in modern homes.

Plants transpire, a process analogous to perspiration in human beings. When we perspire, water is drawn out of our bodies and given off by our skins in an effort to stabilize our body temperatures. Plants are accustomed to climates with sufficiently high relative humidities so that their surfaces are never dry. Most kinds of plants cannot properly function in air that is consistently dry. In an attempt to increase the level of humidity in their vicinity, plants transpire. That is, they draw water out of themselves and release it to increase the moistness of the air.

Even in a hydroponic unit, the supply of water is not inexhaustible. At some point, the plants will have drawn out of themselves as much water as can be gotten—and at that point they will be in trouble. Lacking water, the plant cannot "work," cannot carry out the many processes by which it manufactures the energy to continue growing. As its reserves of energy are used up, the plant will begin to droop, to fail, to become parched looking. In severe cases, in which a rise in the level of relative humidity is not forthcoming, the plant will eventually die.

Modern methods of heating homes have played havoc with humidity, as most of them produce a dry warmth that wrings much of the moisture out of the air. This is not only bad for plants—it's bad for humans too. Some manufacturers have realized this and have begun producing humidifiers which can be hooked up to home heating systems. Other less expensive methods of raising the relative humidity in your rooms can be easily put into practice, and we'll discuss them later on.

But before you decide on what corrective measures are necessary, you have to know how bad the problem is. A hygrometer will tell you. The hygrometer is an instrument designed to measure the amount of water vapor in a location. It is small, efficient and inexpensive—a serviceable hygrometer can be purchased for as little as four dollars. Garden centers, scientific supply houses, and a variety of general merchandise stores stock hygrometers.

You may want to use it at first to measure each of your rooms. Once you've gotten the readings (which may shock you—some homes record readings in winter as low as eight percent), it would be a good idea to keep the hygrometer in a permanent location close to your hydroponic unit. In this way, you can monitor the success of your efforts to raise the relative humidity and also be quickly alerted should the humidity level suddenly drop.

Reading the light levels

Light is an essential part of the photosynthetic process by which plants manufacture the energy necessary to fuel their growth. Because photosynthesis ceases when there is an insufficient supply of light, all plants need a location in which they receive light every day. But that doesn't mean your unit must be built around a window having southern exposure. Different species of plants require different amounts of light. Some plants require only indirect light. Others need many hours of direct, brilliant sunshine. To determine just how much light is falling on your rooms, you'll have to take readings of the levels of illumination, using a light meter.

Light meters are hand-held instruments used in photography to measure the amount of light falling on a surface. These readings are given in units of footcandles—one footcandle is the amount of light falling on one square foot of a surface located a foot away from a burning candle. On a sunny midsummer day, a window having a southern exposure would give a reading in the range of 3,000 to 8,000 footcandles. A window having a northern exposure, checked at the same time, could have a reading of as low as 250 footcandles.

Meters can be costly, so you would do well to try and borrow one from a friend having an interest in photography. If you can't borrow a meter, perhaps you'll have better luck finding a friend with a metered camera—many cameras now have light meters built right in. Or perhaps, should such a friend not be forthcoming, you might want to consider purchasing a camera. Photographs are an excellent way of maintaining a record of your accomplishments at hydroponic gardening. They also preserve your work so that you can turn back to past projects and review the methods you used and the results of those methods.

To take a light reading using either a camera with a meter or a hand-held light meter, you must hold the meter so that its light cell is pointed towards the location to be "read." A window receiving southern exposure need not be measured—you can take it for granted that such a site would provide an optimum amount of light for plants. But any other window, and any other site you have in mind, would benefit from a reading. Hold the meter about a foot away from the site, angled so that the brightest source of illumination falls upon the photo cell. Take readings at intervals throughout the day. You need to know not only how much light reaches a given site, but how long that light is present.

Some light meters have a scale set in footcandles, but most do not, so you'll have to carry through a simple transposition. Set the shutter stop at f4. The shutter speed indicated opposite f4 during the reading would stand for the footcandles. A reading of 500 would thus mean that 500 footcandles of illumination are presently falling on the site being "read."

Correcting improper situations

Once you have determined the levels of light and humidity and the range of temperatures in your rooms, you can decide whether to match plants to the prevailing conditions or alter the conditions so that a greater variety of plants can be grown. While you may find that the amount of light falling in your rooms is sufficient, and even that the temperature range is acceptable to the greatest number of plants, it is unlikely that all three necessities will be present in sufficient quantities. Quite frequently, the relative humidity is the problem.

The following section covers the ways in which you can improve these conditions around plants. A list of books having specific information on the light and temperature needs of various species of plants is included in the Bibliography.

Altering the relative humidity

You have already partially overcome the humidity problem when you raise plants in a hydroponic unit. The regular infusion of water also serves to raise the relative humidity around the plants since some of the water pumped in will rapidly evaporate, becoming vapor suspended in the air. In addition, gathering a number of plants together in a small space will automatically improve the humidity level. Plants transpire at a very rapid rate only when suddenly exposed to dry air. Normally they carry on a much slower rate of transpiration. And the effect of a number of plants gathered close together and regularly carrying out transpiration is an appreciable rise in the relative humidity. One of the reasons tropical rain forests are so very humid is that this process of transpiration is carried out on a grand scale.

Other means can be used to supplement the effect of the water pumped into the hydroponic unit and the transpiration of the plants being grown in it. Bowls of water placed close to heat vents or radiators and regularly refilled will help to moisten the air. If you own your own home, you might want to consider installing a humidifier as part of your heating system. While this addition would cost several hundred dollars, it would surely improve your family's health, as well as the health of your plants. More modest but equally effective would be a portable humidifier which could be carried from room to room as it is needed. If you have a hydroponic unit, you might want to consider adding plants grown only in water to the room. Containers filled with water will certainly benefit the humidity, and the plants in the containers will provide a pleasing contrast to those grouped together in a hydroponic unit.

It is more difficult to raise the relative humidity in a room than to regulate the temperature or increase the amount of light because the procedures for raising the relative humidity are rather inexact. Try to get the level of humidity as close to 40 percent as you can, and then don't worry about it— unless the plants start showing signs of distress.

It is a rare occurrence for the humidity in a room to climb very high. The problem occurs, if at all, only during very warm, very damp days. If you do get terribly high readings (60 percent or over) for several days in a row, increase the ventilation in the rooms in which plants are being grown. Open all the windows. Use a fan. Keep air circulating, but don't let any drafts blow directly across the plants. High humidity isn't directly damaging to plants, but it does provide conditions favorable to the growth of molds or mildews.

Regulating the temperature

Even in well-insulated rooms, there are marked variations in temperature between different parts of the room and at different times of the day. Areas around windows are often prone to drafts. Because heat rises, those areas closest to the ceiling are often warmer than the space extending several feet above the floor. It is probably impossible to enforce a uniform temperature in a room. Instead, it is far easier and wiser to take advantage of these variations, to adapt them to your needs.

Only the hardiest plants should be grown in containers placed on or near the floor. Those plants with the greatest need for warmth (such as ferns) could be suspended in hanging baskets from the ceiling or from a point in one of the walls at least five feet above the floor. If you intend to rely on natural light, your hydroponic unit should only be placed next to a window after you have inspected it for drafts and plugged any cracks that you have found.

If you intend to raise the plants in your unit under artificial light, you are free to select the spot in your room which has the most uniform temperature. The unit should not be placed directly above or in close proximity to a heat outlet. Warm, dry air blowing directly over the plants could trigger a damagingly high rate of transpiration. In addition, the contrast between the very warm air blowing over the plants and the much cooler surrounding air will only confuse the plants' systems, causing uneven growth. This repetitious fluctuation of temperature might even cause the plants to begin dropping their leaves and cease growing altogether.

Most plants are fairly adaptable about the temperature range. They have to be, as even in the warmest of natural environments there are occasional fluctuations in temperature. If you can provide a temperature anywhere within the range of 70 to 75 °F., even the most delicate of plants should prosper.

The growth processes of a plant are dependent on the temperature of its environment. Warmth is perceived by a plant as a signal to go on producing energy by carrying out the process of photosynthesis. During the day, when natural or artificial light is available, warmth is beneficial and necessary. But after the sun has gone down, or after the fluorescent tubes have been turned off, the temperature should decline as well. If it does not, the plant may take the continuing warmth as a signal to keep on generating energy. Night is a time of assimilation for a plant. It is not only essential that plants have a sufficiently high temperature during the day, they must also experience a distinct fall in temperature of at least 10 °F. at night. While it might seem that if warm temperatures during the day are vital to a plant, prolonging such temperatures through the night must spur growth, this is not so.

There is another, allied problem. Plants can adapt to spending the summer in an air-conditioned room if the air conditioner is run throughout the day and night, or if it is run only at night. Plants kept in a room in which an air conditioner is turned on for the day and off for the night (as is usually the case in office buildings) will have their patterns of growth scrambled, interpreting the period of coolness as a time to assimilate energy and the warm hours as a time to produce it. Unfortunately, energy cannot be manufactured without light. The result of this reversal will be the exhaustion of the plant. It may begin to grow sporadically or to shed its leaves and blossoms. It may stop growing altogether. It may even die.

Improving the light supply

It is as rare to suffer from having too much light as it is to be confronted with a dangerously high level of humidity. Instead, most of us face the opposite problem—we lack sufficient light, or we lack the kind of bright light most plants prefer. Unquestionably, your hydroponic unit should go in the brightest location in your rooms. If you're growing plants in a number of water-filled containers, they should be placed in those spots receiving the most hours of bright, indirect light.

You may discover some interesting facts about the light in your rooms when you take readings of the light levels. The supply of light in any room fluctuates greatly throughout the day, waxing and waning as the sun moves across the sky. Windows that are brilliantly illuminated for several hours may be plunged quite quickly into shadow. Windows that you'd assumed to be only dully illuminated turn out, upon the light level being read, to be the recipients of rather bright, if indirect, light. Of course, a condition quite the opposite also often occurs, as a window you'd always considered to have good illumination proves upon inspection to have a very low light level.

Part of the problem in judging the light levels of windows by simply looking them over has to do with our eyes—they adjust to the level of light in a room so that even in a dimly lit location we can quickly perceive the features of the place. Plants can make no such adaptations. It is also certain that most of us don't have any reason to spend any amount of time estimating how much natural light illuminates our rooms. Until we start raising plants, it usually doesn't matter that much. Then it matters a great deal. So don't guess. Take readings to get specific, accurate measurements of the light levels.

A number of foliage plants will do well even in poor light, and you might want to raise some of these species in such spots, placing them in waterfilled containers. All flowering plants, food crops, and seedlings prefer very bright light. The best location for a small hydroponic unit would be close to a window having bright, indirect illumination throughout the day. If possible, it would be preferable to place the unit by a window receiving sunlight in the morning. It isn't sufficient, of course, that the window be brightly illuminated—the light must reach far enough into the room to fall on the plants.

There are several simple steps you can take to boost the supply of light reaching the room. Cut back the branches of any bushes partially obscuring a window. Paint the interior of the window frame and windowsill any light, bright color to intensify the reflection of the sunlight. You might even want to use aluminum foil or a highly reflective wallpaper to line the frame and thus pass on more of the sun's rays to the plants.

If no window receives either bright sunlight or bright indirect light, or if you prefer not to locate the unit close to a window, there is only one alternative—the use of artificial light to replace the absent sunlight.

Hydroponic gardening under lights

Sufficient sunlight to raise plants indoors is a precious commodity. Most of us have, at best, only a few windows receiving enough sunlight for long enough each day to make a suitable location for growing plants. One of the reasons gardeners turn to greenhouses is to escape the frustration of never having enough sunny spots to grow all of the plants they want. There is, however, a less expensive remedy. You can use artificial lights to transform your rooms, to make even the dimmest, most unpromising spots bright enough to support a multitude of plants.

But just any kind of light won't do. Incandescent bulbs, those found in most lamps and fixtures, are not useful. They don't reproduce the full spectrum of light available in natural sunlight; and, without receiving that spectrum, plants can't manufacture food from the nutrients you give them. Fortunately, indoor gardeners now can choose from a wide variety of lights designed specifically for gardening. There are fluorescent tubes designed for plants, as well as metal halide high-pressure sodium (HP) and high-intensity discharge (HID) lamps. They come in various sizes.

First you must choose the size and nature of your hydroponic unit. Then you can match the lighting unit to it. While foliage plants can be grown under regular fluorescent tubes in a pinch, flowering plants and vegetables require much better light. If you are gardening in a room that receives some direct sunlight, you could supplement that supply with fluorescent, HP, or HID lamps. Many of the manufacturers of these lamps supply material to help judge precisely how large a lighting unit you may need, based on the natural light available and the varieties of plants in your hydroponic unit (see Resources, p. 108).

The cost of energy shows no signs of leveling off. To keep your electricity bill from getting out of hand, you can use your artificial lights for more than one purpose. If you're using the light as a supplement to sunlight, gather a number of plants around it. If you must use only artificial light to raise plants, then consider placing your unit in a location that most needs light and color.

To increase levels of light, attach aluminum or metal reflectors to the light fixtures so that all of the light is reflected downward onto the plants. Paint the walls in the room a bright color, so that light will be reflected back onto your hydroponic unit. Or paper the walls in a reflective material.

Keep the lighting units free of dust. Clean the reflectors regularly, *after* removing them from the fixtures.

Hook up an automatic timer to each lighting unit. This way, if you forget to turn on the lights, or when you're not there, the plants will receive as many hours of light as they need.

Restrict the space directly under the lights to those plants that must have intense illumination. Place hardier plants in other areas of a room: they'll still benefit from the lighting units, and they'll give the room diversity.

Many of the units designed for raising plants indoors feature exact designs. *Don't* mix and match parts (such as ballasts) on a lighting unit; always use only what the manufacturer supplies and suggests.

HOUSEKEEPING

Even the most thoroughly automated hydroponic unit could not function for very long without our attention. While hydroponic systems decrease the number of tasks associated with gardening, they do not excuse us altogether. Even in a highly automated unit, there will still be things that we must do. Besides this necessary work, it is advisable that any unit be checked once or twice a day to make sure everything is functioning properly. Plants grown outdoors in soil can be ignored and still continue to secure the elements they need to survive, but plants in a hydroponic unit are entirely dependent on the artificial infusion of water and nutrient salts. They can't rely on rain or on nutrients available in the soil to see them through.

When something goes wrong in a hydroponic unit, you'll find out about it first from the appearance of the plants. Plants that droop, that drop their leaves and the buds of flowers; plants that turn pale or develop patches of pulpy tissue; plants whose leaves have a blotchy or mottled appearance, all carry the same message—trouble. If the problem is not soon identified and corrected, these plants will also be the most visible proof of your error when they collapse and die.

No matter how simple your unit is, check it at least once every day. Make certain that all of the machinery is functioning. Inspect the reservoir to be sure that there is sufficient solution percolating through the unit. Most important, take a close look at the plants. They are variable organisms and their needs may alter from one day to the next. You must learn to tell from their appearance whether they are in need of some corrective measures.

This matter of inspection should not prove as much a duty as it sounds. It has been my experience that most gardeners *want* to spend time around plants. An inspection simply provides a good excuse to do so. As to the matter of plant maladies, a list of symptoms and an identification of the causes of various problems are given in chapter five along with remedies for the troubles. Correctly reading the appearance of a plant to determine whether or not it is in trouble is something that comes easily with experience.

What else should your inspection include besides the appearance of the plants? Check all of the working parts of the unit to be certain that they are functioning correctly. Check the growing medium for any signs of water-logging (water bubbles or puddles of water on the surface) or for the presence of pests or molds.

Inspect the area around the unit and all of the surfaces of the unit as well for any traces of dirt, dust, or other organic refuse. Part of your daily inspection should include any tidying up that is required. It is essential that you keep the surface of the growing medium and the surfaces of all the parts of your unit free of dust, spilled growing medium, and such organic debris as fallen leaves or other plant parts. Dust and debris may attract pests or encourage the proliferation of a disease. Some kinds of flies, drawn by decaying organic matter, can serve as the carriers of plant diseases.

James Sholto Douglas suggests that there should be no smoking in or around hydroponic units. This applies as strongly to small units as to large greenhouses. He notes that "Virus diseases are often carried on dry tobacco leaf. A smoker's hands and face become contaminated, with the result that even an accidental touch may bring about the infection of susceptible plants, such as tomatoes. Cases have been known where viruses have remained in dry tobacco for up to sixty years" *(Advanced Guide to Hydroponics,* p. 227).

Some commercial firms follow very stringent procedures. Workers dress in clean coveralls before entering the greenhouse. Mats doused with Captan or some other fungicide are placed at the entrance to the greenhouses, and everyone intending to enter the unit must wipe their feet to insure that no active disease spores are inadvertently tracked inside. It is also a common procedure to seal off commercial units and fumigate them with a powerful insecticide.

Such serious, elaborate methods should not concern the home gardener. But while you need not emulate the procedures used by commercial growers, you must copy their seriousness of purpose. *Keep the unit clean.* If you cannot eliminate cigarette smoking in the area, do try to keep it to a minimum. And don't neglect your daily inspection. It needn't take much time. It certainly won't take much effort. And you *must* do it. Ignore it, and you begin to gamble with the success of your plants.

Before each new planting your growing medium must be sterilized. Clear out the roots of all the plants you've harvested. This procedure is instructive: you'll have the chance to see just how delicate *and* prolific a plants root system is. Check all of the drainage outlets in the unit. Debris such as root fragments and deteriorated growing medium can quickly gather there and create a drainage problem.

To sterilize the unit, flush it with a solution composed of one cup laundry bleach per five gallons water. Flood the system, blocking all drainage holes. Let the solution stand for a half hour. Drain it off, then flush the system (both your unit and any hoses) with clean water for at least a half hour, running the water continually.

Care in absentia

If the unit should be checked every day, what are you supposed to do when you go on vacation or are absent for several days? The simplest but most expensive method of circumventing the problem is to automate all of the vital functions so that solution is poured into the unit on a regular schedule. If you use artificial lights, a timer could be attached to the fixture.

If your working hydroponics is confined to raising plants in water, you can leave your collection with confidence that nothing can go wrong. Starting about a week before you plan to go, begin changing the water in a few containers each day. Mix up a fresh batch of solution to add to the water. When you return, you should find the plants as healthy as when you left. Hydroponic units using sand or some aggregate as a growing medium require more elaborate efforts to insure the health of the plants.

A small hydroponic unit could be placed in a waterproof container several inches larger in all its dimensions than the unit. After all of the plugs covering drainage holes in the unit are removed, nutrient solution should be poured into the larger container. The solution should come to a level just *below* the edge of the hydroponic unit. It should not be allowed to spill over onto the growing medium. The idea is to allow the solution to be absorbed, drawn upwards through the drainage holes by capillary action. When you return, remove the unit, discard any remaining solution, but *do not* replace the plugs for the drainage holes. Keep them off for one or two days to allow any further solution to drain away and to allow the growing medium to become as thoroughly aerated as possible.

The wick method of watering could be adapted to larger units. Large containers of solution could be placed close to the units, at a point below them. Wicks could then be run from the containers up into the growing medium. Capillary attraction will serve to draw solution up through the wicks into the growing medium. This will suffice to keep the medium moist, but I doubt whether it will provide sufficient solution to satisfy the plants' need for nutrients. However, if you use enough wicks this method can be employed for a few days. To use it for anything more than five days would, in my opinion, be pushing it.

Better than these methods would be recruiting someone to check the unit daily. It would not take long to explain the fundamental procedures, and having someone there to make any necessary adjustments and deal with any unforeseen problems would certainly be preferable to any other method. However, if this is not possible, most kinds of plants are sufficiently adaptable to survive in less than ideal conditions for a week.

Keeping records

It helps to keep a record of your work when you are gardening, and it is essential to do so when you are gardening hydroponically. I suggest that you keep a notebook close to your unit where it is readily at hand. Note the days on which seeds or seedlings are planted, the schedule on which nutrient solution is applied, and the formula for the solution being used. Record any problems you encounter. Keep a count of the number of days required for the vegetable crop to mature, or for the foliage or flowering crop to grow. This is not a make-work project. By examining such records, you'll be able to determine the efficiency of the solution being used. You'll be better able to deal with problems in the future by recording how you dealt with them in the past. And you'll be better able to learn where you've made a mistake and where you did just the right thing. Over the course of several years, your notebooks could become a gold mine of practical information.

There's something else that you might find helpful when you start out gardening hydroponically—a list of basic procedures in the order in which they are to be followed. When I set up my first unit, I found it much easier to have a typed sheet of procedures before me than to constantly be referring back to books, thumbing from page to page in search of the section I needed. The sheet served to remind me of each thing to be done. Once reminded, I remembered each of the procedures, having read enough about them until I was confident about carrying them out. Having the sheet there simply made it impossible for me to skip a step or neglect some regular procedure.

GREENHOUSES

Control is the goal of hydroponics. The hydroponic method allows you to assume greater control over the unreliable functionings of soil and climate. Keeping this in mind, it is not surprising that hydroponics and greenhouses have been closely associated, since greenhouses are a much earlier attempt to assume greater control over the growth of plants. The greenhouse makes it possible for a gardener to regulate the temperature, the level of humidity, and the amount of light available to the plants—to maximize these conditions for the growth of his crops, whether they be of flowers, vegetables, or house plants. The greenhouse forms, in effect, a complete controlled environment. However, it lacks one form of control—the soil.

Pests and diseases have caused greenhouse gardeners no end of woe.

Once they have been introduced into such a controlled environment, in which there is an abundant supply of food and no natural controls (such as predators) to restrain them, they spread rapidly and have a devastating effect on the plants. By replacing the conventional benches in a greenhouse with hydroponic units, you gain control over such problems, eliminating the medium in which pests and diseases flourish. You can exert almost complete control over the plants, supplying them with maximum amounts of the elements they require while greatly reducing the risks usually attendant upon growing plants.

One of the most attractive aspects of hydroponics is its versatility. Its principles can be adapted to use in a unit made of a simple container fitted onto a windowsill, and can be used with equal profitability to run an entire greenhouse. Indeed, all of the largest hydroponic units are to be found in greenhouses, the majority of them being owned and operated by firms producing either flowers or vegetables.

There is an added feature that some gardeners consider of prime importance: the hydroponic greenhouse lends itself to automation. It is possible to install gadgets that monitor and supply all of the plants' needs, to the degree and at that time when they are needed. You need spend as little as fifteen minutes a day working in the unit—the rest of your time can be given over to admiring your handiwork or to pursuing some other gardening interest. It has always seemed to me that in a fully automated greenhouse the gardener becomes a necessary but minor part of the machinery, there to replace fuses, drop in seeds, remove the harvests, and pay the energy bills.

Running an automated hydroponic greenhouse will not be inexpensive. However, you needn't automate everything. Indeed, if you have a strict budget which simply doesn't allow for much gadgetry, you can *still* have a hydroponic greenhouse and keep your initial investment under a hundred dollars.

When do you need a hydroponic greenhouse? When your hydroponic units and your enthusiasm have overflowed your house. Or when you have become serious about supplying much of your family's need for vegetables yourself. Or if, as often happens, you have become so enamored of plants that they take up all of your free space, and much of your free time. When you need more room and more control than you get inside, you must obviously turn to the outdoors. But not to the soil—not, at least, if you wish to continue hydroponic culture.

The greenhouse is your only logical alternative. It should not be considered an alternative reserved for those with land. Greenhouses are becoming a more and more familiar sight in the city, as urban gardeners use their ingenuity to circumvent the problems of space. I've seen porches turned into creditable greenhouses. I've seen large windows outfitted as greenhouses. And I've seen greenhouses in a variety of unusual shapes and sizes sprout on rooftops.

You don't have to be rich to build a hydroponic greenhouse, or to run it. And you don't have to live in the suburbs, or in a rural area. You don't even have to be a master builder. Simple greenhouses require only the most basic tools and techniques to build from scratch. In addition, there are some three dozen manufacturers of prefabricated greenhouses offering an incredibly wide range of sizes and styles, with prices ranging from very modest to very expensive.

Another alternative, while unlikely, is worth considering. As the market in plants has altered, and as fortunes in the horticultural trade have risen or fallen, commercial greenhouses have been shut down or abandoned by their owners. I live in an area that is an uneasy mix of suburban and rural influences. Within a fifteen-minute drive of my home are three greenhouses. While none are outwardly in very good shape, they seem to be structurally sound. If I knew of a group interested in expanding into greenhouse gardening but not yet ready or willing to erect one for themselves, I'd suggest they pool their funds and labor to acquire and renovate such a place. I don't want to imply that such opportunities occur frequently, but they do occasionally. Even in cities there are sometimes such greenhouses to be found, relics of an earlier time, often indications of an alteration in the composition of a neighborhood. How much it would cost to renovate such a greenhouse depends entirely on its condition, but even one in which most of the panes of glass have been punched out and the interior given over to dust, rust, and rats, can be salvaged if the structure itself is sound.

Greenhouses have, for the greater part of their history, been known as the exclusive possessions of the wealthy. When they first assumed their now-familiar shape—a structure longer than it is wide, with a slanted roof— some four hundred years ago, they were quite literally the preserves of kings. It took some maneuvering before even the nobility were allowed to build greenhouses on their estates. For a long time they were known as "glass houses" because they were composed almost entirely of panes of glass. Indeed, greenhouse construction was not possible until the science of producing large sheets of glass had been mastered.

In much the same way, it took new developments in technology to make an inexpensive mass-produced greenhouse possible. Until the advent of lightweight materials such as aluminum, polyvinyl chloride, and polyethylene, the materials needed to build a greenhouse were expensive and could be used only by someone with a good deal of experience in construction. The light synthetic products released after World War II revolutionized the building process. These new materials could be worked easily, even by someone with little experience. They were inexpensive. They were so light that they could be shipped through the mails in precut pieces, ready for assembly. Not only were greenhouses suddenly within many more people's budgets—they could now be mass produced and sold nationally. The development of the modern greenhouse, and especially of the prefabricated greenhouse sold as a kit, brought a more sophisticated form of gardening within the reach of many more people. It "democratized" the greenhouse. Hydroponics can be seen as the most recent extension of this process, bringing more and more sophisticated gardening methods, tools, and techniques within the budgets of larger and larger numbers of people.

Building vs. buying

Should you build your greenhouse from scratch or buy a prefabricated unit? Many of the texts on greenhouses dismiss the question with a suggestion that readers buy a unit that can be easily assembled. Admittedly, there are good reasons for doing so, but that's not the whole story.

Several dozen firms now manufacture and sell prefabricated greenhouses. These units cover a wide range of styles, sizes, and prices. Many of the units can be assembled in a weekend using simple tools. The kits include precut pieces of wood and metal to form the frame of the greenhouse, and glass, plastic or fiberglass in premeasured sheets or rolls to be used as the covering for the unit and to hold frame and covering together. Kits for small walk-in greenhouses can be found for as little as $150. Larger prefabricated units may cost as much as $1,000. Several firms even design greenhouses to fit individual specifications. Such custom greenhouses are planned and constructed by the company at a sizeable cost. And then there are commercial units, sold to those making a living from gardening. Such units cost anywhere from $15,000 *on up*—and that's without much of the operating equipment.

This raises an important point. Even if you want no more than the least-expensive prefabricated unit, you'll still have to add the cost of outfitting it. You'll need benches or some other form of container to hold your plants. You'll need a heater if you intend to run the unit through the cold months. You'll need a fan to properly ventilate the unit. And you'll need yet other, though less costly, accessories. If you want to lay a concrete foundation for the greenhouse, you'll face yet another expense.

Of course, you'll need all this equipment whether you build or buy a unit. But you shouldn't assume that a kit is necessarily going to cost you less money or time than building a unit from scratch. The Bibliography lists other books which include plans for do-it-yourself greenhouses, and the Resources section gives firms selling plans.

A very simple greenhouse can be built by anyone having a few necessary tools and the time to use them. Such a greenhouse is good enough to get you started, to allow you to find out if greenhouse gardening is something you want to keep on doing. Then, as your confidence and your aims expand, you can investigate buying or building a more permanent unit. Building a spacious, permanent greenhouse is not a simple task, and if you are inexperienced in working from plans and using a variety of tools, you had better look for an alternative to doing it yourself. Perhaps you have a friend who would be willing to guide your efforts. Perhaps a local handyman could carry out the most difficult procedures for you. Or lacking experience, knowledgeable friends, or local help, you can buy a prefabricated unit.

Before you decide what to do, investigate the possibilities. If you find a plan for a greenhouse that suits you, try pricing the necessary materials locally. Then send away for catalogs and price lists from at least several manufacturers of prefabricated units. Compare the costs of similar units—those you could build and those you could buy. After you've begun to work out the comparative costs, draw up a list of the equipment you must have to make your greenhouse operational. Then, and only then, can you make an informed decision on whether to build or buy, and can you arrive at an estimate of how much your plans will cost.

When you've come down to the point of settling on one or the other course, don't decide to do it yourself only because it *might* save you money. It might— but it might not. Don't try it yourself unless you know how to do it. If you jump into the project without being prepared, you'll likely end up spending more than you'd counted on to rectify mistakes. I have one friend who gave up on his greenhouse when it was half completed. At that point, he discovered he no longer knew how to proceed. It took several weeks of brooding and several more weeks of reading before he was ready to go back and get the job done. I remember meeting another greenhouse gardener who, after I had admired his unit, ruefully admitted that he had had to call in professional builders to set right a number of things that he had gotten wrong.

Types of greenhouses

The term "greenhouse," while being both admirably imaginative and accurate in communicating the purpose of such a structure, is misleading on the matter of size. greenhouse has come to refer to any structure having as its purpose the control of the environment to better produce healthy plants. There are window greenhouses, free-standing cabinet greenhouses, and even greenhouses small enough to be displayed on a table. All of these units fit within one's rooms and cannot properly be referred to as "houses." Yet until some better term comes along, that is what they will be. So don't immediately assume that a greenhouse is only within the means of someone with plenty of money and plenty of land. Even if you live in an apartment you can have a greenhouse. Even if your budget is very tight you can afford some sort of greenhouse.

The largest greenhouses sold as prefabricated units and featured in plans for those who prefer to build their own are known as free-standing units. The old glass house shape is still quite popular. Other recent introductions include the Quonset hut style, the "gothic" greenhouse, and the geodesic dome greenhouse.

Lean-to greenhouses can be almost as large in total floor space as a free-standing unit but differ in that they are designed to be built against an existing structure, such as a wall of your home. Lean-to units generally have two short end walls and one long side wall facing outwards. Some have a door in one of the end walls. Of course there are variations in this design. Some homeowners cut an opening through the house wall leading into the lean-to, integrating it more strongly into the structure of the house. A recent variation on the idea of the lean-to unit is the sun pit, a half-dome built against the wall of an existing building.

Then there are the smaller greenhouses—window greenhouses or units built to be displayed indoors. And there are the semi-greenhouses, such as "sun rooms" filled with plants, their environment modified by the addition of a humidifier, a fan, or even a small heater.

Selecting a site

Your greenhouse must be erected on the sunniest piece of land you have. The point of having a greenhouse is to further extend your control of such uncertain elements as temperature and humidity, while maximizing the supply of light available to the plants. You can't do that if trees or nearby buildings cast shadows over the greenhouse, depriving your plants of several hours of sunlight every day. The best—indeed the only—location for a greenhouse is on a plot of ground receiving direct sunlight throughout the entire day. Of course, you could add fluorescent tubes to the unit to compensate for the lack of natural light, but what would be the point? You're supposed to be tapping into a free source of light to avoid any increase in your utility bill.

Once you have some idea of the size of the greenhouse you intend to buy or build, take a careful look at your land. Look at it in the morning, in the early afternoon, and in the late afternoon. The spot never obscured by shadows, or obscured for the least amount of time, is the spot you want.

You'll need a plot that is relatively flat. Ground that is very uneven will require hard work and plenty of landfill to even out. Before you start laying out the foundation, make certain that no pipes or utility lines lie beneath your chosen site. Should these lines have to be excavated at a later date, your greenhouse will have to go. If the only site available to you is on a hillside, use it—but only if you can build on the *south* side of the hill. Unless you have no choice, don't build there at all. Cold air, being heavier than warm air, will collect first in the lowest spots in a landscape. This could pose a sufficiently serious threat to your plants to compel you to put a heater in the greenhouse. And that means a higher utility bill. In addition, there is the potential for a problem with proper drainage, a possibility when building close to any slope.

The further away your greenhouse is located from your house, the more costly it will become to run utility lines to it. If you intend to build a large free-standing unit, you will have to have power. You'll need electricity to power a fan and a heater, and perhaps heating coils and a humidifier as well. You'll need water. Though you may be able to make do by carrying it from a nearby outlet, this is sure to prove a tiresome procedure. If you're going to build a large unit, it's better to go all the way with it, to put in everything you'll need. To do otherwise would be to compromise your efforts.

Of course, such costs of installation are much lower when you build a lean-to greenhouse against the side of your home. And you should really only build it against the wall receiving eastern or southern exposure. Placed anywhere else, the unit will require fluorescent tubes for supplemental lighting. Even then, you may be unable to raise some of the more delicate, light-demanding plants. So build along the southern wall of your home, or the eastern if the land along the southern side is unacceptable.

4. What To Grow in Your Hydroponic Garden

It would be easier to name the plants that cannot be grown using hydroponic methods than to name those that can. Indeed, the very great list of possibilities is largely limited only by the ease with which the plants can be grown. For instance, it would certainly be possible to grow large shrubs and even trees in a hydroponic unit, but it would not be practical. You could, however, start such large plants in a hydroponic unit to better protect them and insure their rapid development, transplanting them when they become sufficiently sturdy. You can grow most kinds of house plants, many kinds of annual flowering plants, most kinds of vegetables, and many sorts of herbs indoors in a hydroponic unit.

As mentioned previously, the requirements of a plant are not greatly altered when it is grown by hydroponic methods. It still needs sufficient levels of sunlight, moisture, and warmth. The only other point to keep in mind when selecting plants is to determine if the species has a preference for "light," airy soil or for "heavy," substantial soil. Plants preferring a "light" soil tend to do better when placed in a growing medium like perlite that allows the unhindered circulation of air, as opposed to a somewhat more compacted growing medium such as sand.

It would require a book the length of this introductory text to cover all the plants that can be grown hydroponically. This chapter is intended as a brief overview of the possibilities. The Bibliography includes a list of books having specific cultural information on each of the kinds of plants mentioned in this chapter. When you are ready to begin gardening hydroponically, you should consult such books to determine what kinds of environmental conditions your plants need and also to establish any other regular procedures, such as pruning, that they might require.

FLOWERING AND FOLIAGE PLANTS

The hydroponic process lends itself to a variety of uses for the gardener raising foliage or flowering plants. A unit can be used to raise seedlings, and as the plants mature they can be moved to other hydroponic units in your home to be displayed. Or, once they have reached a sufficient size, they can be moved outdoors to your garden. If you haven't enough indoor hydroponic units for all your plants, you can transplant them into soil-filled containers, or you can raise many plants in a large unit and then distribute them among several smaller containers outfitted for hydroponics.

You can use a hydroponic unit to bring plants into flower outside of their normal season. In fact, this is the most frequent use of hydroponics in commercial floriculture. English "flower factories" churn out carnations, and in the United States huge hydroponic greenhouses are maintained so efficiently that several thousand bunches of flawless flowers can be produced every week. In some parts of the country, it has become more and more likely that the flowers you admire in the local shop came from a hydroponic greenhouse.

I can think of only one difficulty associated with the hydroponic cultivation of nonfood crops—you may end up with many more flowers and plants of a particular species than you desire or have any use for. Because each kind of plant has its own particular set of requirements, many varieties can be grown only in containers holding no other species. For instance, roses need a greater amount of potassium in their nutrient solution that most other plants need or can stand. Species also differ in the level of light, heat and relative humidity they prefer—and these qualities cannot easily be compromised without damaging all the plants in a unit. This means that you must be very careful in matching species to share space in a unit, or you must resign yourself to having very large crops of some plants.

You can give them away, of course. Or you could try growing a number of different varieties of one species together, such as raising many kinds of begonias in one unit. While the begonias occur in an astonishing range of leaf patterns and colors, many varieties share basic requirements of light and warmth. Another alternative is to do all your indoor gardening in small hydroponic units so that you can have great variety without having masses of plants.

One enthusiastic hydroponic gardener I know maintained only a single large unit. He used it to produce seedlings, which were then transplanted to individual pots filled with aggregate. After the seedlings had been cleaned out, he used the unit to produce a crop of flowering plants, which were then transplanted outdoors. Next he started his tomato plants in the unit. Within the course of a year, he used the unit for a variety of purposes, while consolidating many diverse gardening functions.

A plant's needs do not alter according to the material in which it is grown. A plant being raised in a hydroponic medium will still require the same light level, temperature range, and relative humidity as if it were being grown in soil. Thus, when raising any plant in a hydroponic unit, you must follow the cultural needs of the plant just as you would if you were raising it in a pot full of soil or outdoors in the ground. Any thorough gardening text should contain the information you require. Books on these subjects that I consider especially helpful are listed in the Bibliography.

GROWING VEGETABLES IN A HYDROPONIC UNIT

Hydroponics holds great promise for developing nations as a means of producing large yields of vegetable crops in less time than is required by conventional farming methods. James Sholto Douglas has summarized his extensive research in the field by stating that "hydroponic vegetables and fruits mature more quickly than those planted in soil, produce far higher yield and need less space" *(Beginner's Guide to Hydroponics,* p. 83).

Obviously, this is of great importance to nations chronically short of food, but principles of interest to countries apply with equal relevance to individuals. If you live in a situation in which you lack sufficient land to plant a vegetable patch, even if you have no land at all, you can still raise fresh vegetables. If you have land and prefer to raise vegetables or fruits for your family, you can raise more vegetables faster using a hydroponic greenhouse in place of a plot in your back yard. Even if you have no great interest in the lower cost of raising vegetables or fruit, a vegetable plant or a display of fruit being grown in small hydroponic units makes an arresting sight.

Among the vegetables that can be grown hydroponically are: artichokes, asparagus, beans, beets, broccoli, brussel sprouts, cabbages, carrots, cauliflowers, celery, cucumbers, eggplants, leeks, lettuce, melons (cantaloupes, muskmelons, watermelons), onions, parsnips, peas, potatoes, radishes, rhubarb, spinach, squash, tomatoes, and yams. Berries, such as raspberries and strawberries, can also be grown in hydroponic units. Fruit trees can be started in a unit and transplanted outdoors when they become sufficiently large and robust. Dwarf varieties could even be kept in a unit throughout their productive life span.

Vegetable plants grown in soil must be spaced as much as a foot apart so that their root systems will not tangle and compete for necessary supplies of moisture and nutrients. Because plants grown hydroponically have all the moisture and nutrients they need pumped to them, the roots need not sprawl over any great distance. Thus, plants in a hydroponic unit can be placed in aggregate at half the distance apart recommended for their cultivation in soil.

The basic requirements of care for each crop are not altered by the switch from soil to aggregate. You must give the plants the levels of light and warmth they need, and you must follow all of the normal procedures of caring for the crops. Several books on the cultivation of vegetables which I have found to be particularly helpful are listed in the Bibliography. Only a few matters require emphasis, and they are listed below.

Artichokes. Require larger amounts of potassium than most other plants. A perennial crop, artichokes will not appear in appreciable amounts until the second or third year.

Asparagus. Will not begin to produce until the third year. Requires an aggregate loose in texture and well aerated, also larger than usual amounts of potassium and potash. While it would be interesting to raise artichokes and asparagus hydroponically, I prefer not to do so. Because both are perennial crops, they must be maintained from year to year, and this permanently ties up units that might be more usefully employed.

Beans (broad, French, lima, soy). Bean crops need a loosely textured aggregate, which should never be allowed to become waterlogged. Beans need less nitrogen than other crops (they are nitrogen-fixing plants and actually produce it), but they need more than the usual amount of phosphorus, potassium, and sulphur.

Beets. Calcium, potassium, sodium and chlorine are all important to the growth of beets.

Broccoli, cauliflower, cabbages and *kale.* All members of the Brassica family. They need a well-aerated growing medium and abundant amounts of nitrogen, phosphorus and iron. Don't allow the growing medium to become waterlogged.

Carrots. Short varieties of carrots do best in hydroponic culture. Because the plant develops below ground level, a full-sized variety would require a very thick bed of aggregate. The aggregate should be light and well aerated.

It is important that this crop receive adequate amounts of potassium and phosphorus.

Celery. A well-aerated medium is essential. Sodium and chlorine must be included in the nutrient solution.

Cucumbers. Appreciate a high relative humidity, frequent irrigation, and protection from intense, direct sunlight. Somewhat more demanding of care than other vegetable crops raised hydroponically.

Eggplant. Requires ample amounts of nitrogen, phosphorus and potassium. However, as soon as the fruit on each plant is noticeable and well formed, the amount of nitrogen in the solution should be reduced by one-third, since too much nitrogen at this stage will spur the growth of foliage at the expense of the fruit. These plants will bear several times during a season, so it will be necessary to repeat this procedure each time the developing fruit takes on its characteristic shape. Well-irrigated aggregate is required.

Leeks. Need large amounts of phosphorus, ample amounts of nitrogen and potash.

Lettuce. Especially sensitive to drainage and requires a well-drained, well-aerated medium.

Melons (cantaloupes, muskmelons, watermelons). All require good irrigation, but are vulnerable to damage caused by very high relative humidity. Keep the growing area well ventilated.

Onions. Require ample potassium and nitrogen.

Parsnips. Because the roots of parsnips are exceptionally long, the aggregate must have a greater depth than is normally used. Large amounts of potassium are essential. Calcium, nitrogen and potassium are also important.

Peas. Both dwarf and tall varieties, early and main crop types, can be grown in hydroponic units. The aggregate must be well aerated. Supports, onto which the plants should be trained to climb, will be necessary.

Potatoes. Tubers must be well covered with aggregate or they can be badly damaged. Ample supplies of potassium, phosphorus and iron are required. The pH of the solution should be slightly acidic.

Rhubarb. Requires plenty of phosphorus.

Spinach. Requires well-aerated medium with good drainage. Needs plenty of nitrogen. Take steps to keep the growing medium from becoming waterlogged.

Squash. Requires a medium with good drainage.

Sweet potatoes. Need ample amounts of potassium, calcium, magnesium and phosphorus. Best grown in a sand medium.

Tomatoes. One of the most popular of hydroponic crops, and justly so. Seem to do very well when raised hydroponically, ripening as much as eight weeks earlier and producing more fruit than tomato plants grown in soil. Require a growing medium that drains rapidly and is sufficiently loose to be well aerated. Plants also need as many hours of direct sunlight as possible, temperatures in the seventies, and a relative humidity of from 40 to 50 percent.

Yams. Well suited to sand culture. Require good drainage.

Several methods adopted from gardening practices in soil will prove useful to the hydroponic farmer. You can increase the amount of vegetables you can grow in a bed or other hydroponic unit by *intercropping*—that is, by alternating tall and short plants. Some vegetables take up much more space with their foliage than other crops. Tomatoes are a good example. Lettuce, on the other hand, is a ground plant, growing just above the surface of the aggregate. Tomato plants grown side by side tend to take up a good deal of space. By alternating tomatoes with lettuce or some other low-profile plants, you can fit more crops into a given space.

Catch cropping is the process of sowing seeds that are quick to develop (such as salad greens) among the stalks of larger, slower-growing crops. Before the slower crops are fully developed, the fast plants will have reached maturity and been removed.

You can also save space by growing crops having similar cultural requirements and needing a similar nutrient solution in the same unit. In short, condense your crops as much as you can to save space, to save the time and effort involved in running a number of units, and to save nutrient solution.

HERBS

Herbs mature rapidly, require little care, have a wide variety of uses and seem to do well in hydroponic culture. Even a small hydroponic unit can produce an impressive number of herbs, and several manufacturers have produced kits designed especially for their production.

The number of species and uses of commonly available herbs are sufficiently numerous to require book-length treatment. See the Bibliography for a list of books which will give you the necessary information.

OTHER CROPS

A good many crops can be grown using hydroponic principles, but home gardeners have neither the time, the space, nor the money for them. Among these crops are cocoa, corn, sugar cane, rice, rubber, tea, tobacco, and cereal grains. In many cases, hydroponic units are used to get these crops started, and when the seedlings reach a sufficient height they are transplanted to the fields. Animal feed has also been produced in hydroponic gardens, and specialized units for the mass production of animal fodder are now available.

This provides one instance in which the crop being grown is measurably improved when raised hydroponically. James Sholto Douglas, in *The Advanced Guide to Hydroponics,* notes that "the protein content of grain or seed increased dramatically when it was germinated and grown to form young grass in hydroponic units." He goes on to report that "Numerous farmers in different countries state that the milk yields of their dairy cows have risen substantially after the feeding of hydroponic forage was commenced. The system is especially valuable in arid or semi-arid regions, where there is little or no field grazing, in cold and harsh climates, where stock must be kept inside much of the time, or at times of drought" (p. 284).

The variety of crops which can be profitably or usefully grown in a hydroponic unit is certain to continue to expand as researchers around the world test out more and more crops in experimental hydroponic set-ups. this is certainly of interest to us economically. Such developments might also be of real help in alleviating the food shortages afflicting many developing nations. Many such countries already have hydroponic units of considerable size to turn out vegetables. The use of these units to product animal fodder and grains for themselves, as well as nonedible crops for sale abroad, could be of great help in making these countries more self-sufficient.

GROWING PLANTS FROM SEEDS AND CUTTINGS

If you are serious about gardening, if you want to continue to expand and diversify your collection of plants, the only sensible way to do so is to grow your own. Most plants can be raised without great difficulty from seed, and some kinds of plants can be raised from cuttings. The seeds of many species of plants can be purchased from a variety of firms.

A cutting is a piece removed from a root, stem, or leaf of a mature plant. Placed in a growing medium and carefully tended, this slip will develop into a mature plant of the parent species. Many gardeners regularly trade cuttings to acquire new species of plants. I think this is a wise and economical method of adding to your collection. It also serves to bring you further into the company of fellow gardeners, a group often distinguished by their enthusiasm and generosity in sharing knowledge.

It is neither costly nor particularly difficult to raise healthy plants from seeds or cuttings. Indeed, in some ways, the principles of hydroponics lend themselves to the rearing of plants. Growing mediums are usually of an even texture, light and airy, free of soil-borne diseases, and distinguished by being moist throughout. All of these qualities make aggregates superior to soil for the purpose of germinating seeds.

Damping-off disease is a fungus transmitted in soil. It primarily affects seedlings, attacking the tissues of a plant, causing it to lean woefully to one side. It is almost always fatal, and can destroy a seedling in as little as a day. Damping-off is the most common cause of fatalities among seedlings. When you germinate seeds in a hydroponic unit, it cannot occur. For that reason alone, it would be preferable to start seeds in a hydroponic unit.

Seeds and germination

Seeds are plants in potential. The embryonic plant encased in a hard coat has already been differentiated into several parts—a stem, one or two leaves, and a miniscule root. Each seed also contains a supply of sugars, proteins, fats, and starch to fuel the initial growth of the plant once it has burst free of its protective covering.

However, the plant will not emerge until certain favorable conditions exist. If sufficient water, warmth and oxygen are not available, the plant will not shed its coat. In this manner, seeds can survive for long periods of time until conditions turn in their favor. Once these initial conditions are met, the plant will germinate—that is, it will shake off its protective coat and push up through the growing medium. Once this has occurred, an adequate supply of light will be essential if the plant is to keep on growing.

Although a seed will not germinate until all three elements are present, water is at first the most important, for it softens the hard outer layer of the seed, allowing the plant to emerge more easily, while at the same time causing the embryo to swell outwards from within. To effectively protect the seed, the outermost layer must be very hard, and it requires this combination of the embryo swelling from within while the coat is softened by water for this layer to be cast off. In addition, water serves as the means by which oxygen is carried to the embryo.

Only a small percentage of the seeds generated by plants in natural environments will receive just those conditions they need to germinate and survive. In hydroponic gardening, you can improve on nature. If you supply adequate amounts of warmth, moisture, and air to seeds you've placed in a growing medium, you should be able to consistently produce a very high rate of successful germination.

Plants from seed

If you intend to make a steady practice of raising plants from seed, the most efficient manner in which to do so would be to set aside a hydroponic unit for that purpose alone. Because seeds and seedlings have very specific needs which must be met, it would be impractical and awkward to attempt to mix mature plants and seedlings in a unit.

A word of caution before you begin. When working with seeds, make certain that you don't bring your hands up to your mouth. Many kinds of seeds are now pretreated with a fungicide, and the chemicals could be harmful if you swallow them.

Whether you start seeds in soil or in an aggregate, the procedure is largely the same. After the aggregate has been poured into the container, small holes are made in the surface with the tip of a pencil or some other narrow instrument. These holes should be no more than a quarter of an inch deep. The seeds are tucked into place and aggregate gently brushed over them to fill in the holes entirely. Very small seeds need only a thin layer of aggregate spread over them. Indeed, when dealing with nothing but small seeds, it would be simpler to arrange them on the surface and then spread a thin layer of aggregate on top.

The best way to distribute the seeds in the container is also the most orderly. Arrange the seeds in parallel rows. Seedlings growing in fairly straight rows will prove much easier to thin out and, when the time comes, to remove for transplanting. It's a general practice to sow more seed than you need—some may fail to germinate, while others may grow in rather sickly and must be discarded.

Before you plant any seeds, water the aggregate enough to make it thoroughly moist but not waterlogged. After the seeds have been set into place, give them a *light* watering. Thereafter, the growing medium must be kept constantly moist, but never so moist as to be sodden.

The day after the seeds have been sown, you can substitute nutrient solution for plain water. Thereafter, always apply solution rather than water. The solution should be the same dosage as used for mature plants. You can either allow the unit to drain after each application or you can keep the unit plugged, removing the plugs every two to three days so that any build-up of excess solution can drain out. Unless the weather is very warm, applications of solution every other day should suffice. As the plants mature, more frequent applications of solution will be necessary.

The first two weeks are the most critical time for the seeds. To germinate properly they need warmth as well as steadily moist conditions. Most kinds of seeds require a temperature in excess of 65°F. to germinate. A waterproof insulated heating cable is the best, surest way to guarantee sufficient warmth for the seeds. Such cables are available with or without a thermostat to control the degree of warmth, and in sizes from twelve feet up to one hundred and twenty feet in length. (A cable twelve feet long will heat an area of four square feet.) The cable should be laid along the bottom of the unit, and arranged in a series of rows held in place by staples (or electrical tape, should your unit be made of heavy plastic or some other material which could be damaged by staples). The outermost row of cable should be at least two inches from any side of the box. The cable should not touch the thermostat, nor should lengths of cable cross over one another—in both cases, insufficient heat will be generated. Bring the plug end of the cable out over the rim of your unit. After the cable is firmly in place, fill your unit with whatever growing medium you have selected. Many cables have thermostats preset to 70°F., an excellent temperature for germination.

Your hydroponic germination unit should be shielded from the direct rays of the sun until the seeds have sprouted and sent shoots above the surface of the aggregate. While direct sunlight is harmful, bright but diffuse light is not only acceptable, but essential. After the seedlings have appeared you can give them full exposure to the direct rays of the sun, protecting them only during the midday hours when sunlight is most intense. If the unit is of manageable proportions, it would be helpful to rotate it each day to prevent the seedlings from growing too far in one direction, since they are inclined to bend towards the dominant light source.

After the *second* pair of leaves appear on a seedling, it is ready to be transplanted.

Transplanting seedlings

The second pair of leaves a seedling produces are considered the "true" leaves. Once these have developed, the seedlings can be moved to the permanent location and container you've selected for them. You will find it somewhat easier to remove the seedlings if the growing medium has been allowed to become drier than normal.

Use a spoon to lift each seedling out of the growing medium. Carefully probe through the medium and, gently grasping the seedling by one of its leaves, use the spoon to lift the plant upwards and out. Never grasp a seedling by its stem—it is so fragile that even careful handling could fatally damage it. You should try not to handle the roots either.

Before you lift a seedling out of the container in which it has been started, you should have prepared a suitable hole in the aggregate of its new container. Gently set the seedling in place and brush the aggregate in around the stem and over the roots until the soil line (the point reached by the aggregate in the original unit) has been met. While you're brushing the aggregate in place, continue holding the seedling by a leaf so that it is standing perfectly upright in the unit.

After the container selected has received its full allotment of seedlings, you can apply a full dose of nutrient solution. Should you prefer not to set aside an entire unit for the propagation of plants, there are several alternatives.

Jiffy-7s are small square or round pieces of compressed peat around which a thin plastic net has been wrapped. When the peat is watered it expands until it is fully two inches high, and one and three-quarter inches in diameter. One or two seeds can be tucked just below the surface of the peat, which supplies some nutrients while having excellent water retention qualities. The roots of the seedlings will penetrate the netting easily.

Jiffy-9s are quite similar, except that they do not have plastic netting. This makes them somewhat more likely to crumble when you are working with them.

Fertlcubes are square, one-inch blocks composed of moss, perlite, and vermiculite, with a dose of nutrients mixed in.

Jiffy pots and fertlcubes are easy to handle. You can remove seedlings from such containers when the plants have sufficiently matured or you can place the entire cube in your growing medium, bringing the edge of the block up to the edge of the aggregate. Because they are composed of sterile materials, there is no chance of the containers contaminating your plants. You will, however, have to remove the netting from Jiffy-7s. Aside from that, such containers can easily become a part of a hydroponic unit. Since peat is as absorbent as many mediums, there is no chance that plants left in such containers will receive less moisture or nutrients than other plants in the unit.

Plants can be propagated in a variety of other containers. Large flats, peat pots, and complete enclosed propagation units are all available. As H. Peter Loewer observes in his excellent book, *Seeds and Cuttings,* "Just about any container that will hold the planting mix, that is fairly waterproof, has drainage holes on the bottom and is at least over 2½ inches in height will work for seeds" (p. 34).

5. Preventing and Treating Problems

C.E. Ticquet, in his book *Successful Gardening Without Soil,* stresses that hydroponics is not a method "in which anything can be left to chance. When you are growing in soil, you can assume that if you put a seed into a portion of soil—any soil, anywhere—it will grow. You will be wrong, often, but you will be right often enough to justify the assumption. Nature sees to that. In soilless culture, however, in a way *you* are nature. *You* are providing the substitute for soil. And just as Nature takes care of everything, you must be prepared, in this instance, to do the same" (p. 150).

Throughout this book I have stressed the relative ease with which small hydroponic systems can be built and maintained. But this ease of operation depends on the few necessary actions *always being carried out.* You cannot postpone your duties by even a single day. If you repeatedly miss necessary chores, the plants are sure to suffer. Responsibility is an important quality in the home gardener. It is an essential quality for anyone wishing to succeed, no matter how modestly, in raising plants hydroponically.

I don't mean to imply that if something does go wrong, you should be consumed with guilt. Even the most experienced gardeners make mistakes, forget to carry out some necessary procedure, or do the wrong thing at the wrong time. When anything goes wrong, you must first identify the problem, set about correcting it, and only then should you concern yourself with determining if you have been at fault. Most of your mistakes will be educational. There is no more forceful way to learn to avoid certain difficulties than to have suffered through them. I once brought into my house a plant infested with aphids. Because I was in a hurry to keep an appointment, I neglected to check the plant for signs of pests or diseases. By the time I discovered the infestation, several other plants had been affected. It cost me time and effort to rid my plants of the aphids, and two of the plants had to be discarded. Now I check every plant thoroughly before I bring it into my home. And I keep each newly purchased plant isolated from the rest of my collection for a period of two weeks—long enough for any disease to show itself or for any pests to emerge from the soil.

If the time you can spend maintaining a hydroponic garden is severely limited, automating the unit will assure the plants of all those elements they need to survive and prosper. However, automation will require both an initial outlay to purchase the components and increased energy bills. If you can't afford to automate, or if your unit is too small to make automation worthwhile, you'll have to apply the solution and aerate it yourself.

If your schedule is too erratic to permit you to carry out these functions on a regular basis, you would do better to stick to growing plants in soil or restrict your hydroponic activities to single plants grown in containers of nutrient solution.

Unless you repeatedly neglect the proper maintenance of your unit, the problems you encounter will likely be due to either an imbalance in the system or to pest and disease attack. Your plants will signal you when there is a problem. Leaves that become very pale or turn yellow or brown; leaves growing in distorted, misshapen fashion; wilted leaves; leaves dropping in large numbers; oozy, pulpy plant tissue; and buds and flowers that wither and drop off are all indications of some malign force at work.

When such symptoms occur, you may be hard pressed to identify the cause. I have found that it helps to have a check list of potential difficulties on hand. This way you can establish a precise procedure for isolating the problem, eliminating unlikely sources of trouble as you move down the list.

The number of maladies may seem worrisomely large, and the problems described in this chapter may sound complex and deadly. But unless you're doing something very wrong or are the victim of profoundly bad luck, you should never experience most of these problems firsthand. Don't expect difficulties to crop up every week, or every month. If you maintain your hydroponic unit properly and give the plants those conditions that they require, serious problems should be few and far between.

CHECKLIST FOR TROUBLE

Environmental Conditions

Are the plants receiving sufficient light?
Are they receiving too much light?
Is the unit receiving sufficient ventilation?
Is the temperature in the area very high or very low?
Is a cold draft passing over the plants?
Is the atmosphere very dry?
Do you know what the relative humidity generally is in your rooms or
    greenhouse?
Is the air particularly laden with pollutants?

The Unit

Is the aggregate sodden or waterlogged?
Is the aggregate drying out much faster than is usual?
Is the unit draining too rapidly or too slowly?
In larger units, check the nutrient reservoir to determine if the level has
    fallen too low for the pump to draw it out.

Water

Check the pH level of the water you're using.

Nutrient Deficiencies

Check your nutrient formula to determine whether there might be an
imbalance of the ingredients. Review your procedure in preparing the
formula.

If none of these items seem to be the cause of the difficulty, consult the
tables on pests and diseases (p. 97). For information on pesticides, see p. 102.

SYMPTOMS AND CAUSES OF NUTRIENT DEFICIENCIES

Symptom	Deficiency
Slow growth rate. Leaves lose color, taking on an unnatural light green shade or becoming yellow. Lower leaves affected first.	Nitrogen
Leaves turn very dark shade of green with further blotches of discoloration. Leaves may also turn gray. Underdeveloped root systems.	Phosphorus
Lower leaves become brown with darker blotches. Leaves become dry, curl up, yellow.	Potassium
Leaves fail to fully develop and are abnormally small, dry and dark. Growth stunted, roots underdeveloped.	Calcium
Leaves yellow. Buds fail to develop and bloom. Pale or brown spots on leaves. Leaf veins remain green.	Magnesium
Only veins remain green, while remainder of leaf loses its color, becomes dry, crinkly. Loss begins at tips.	Iron
Buds fail to bloom. Growth rate of plant slows down. Leaves appear mottled in contrasting pattern of dark and light.	Manganese
Veins turn yellow. Section of leaf closest to stem becomes very dark.	Sulphur
Stems torn. Dry, spindly leaves. Brown tips develop.	Boron
Growth rate slows far down or even ceases.	Zinc
Leaves quickly become very brown, dessicated, and drop off. Some plants quickly die.	A severe overdose of nutrients is rare, but can occur. It is almost always fatal if not immediately corrected. However, to be damaging the overdose must be very, very severe. You'd have to do everything but pour in undiluted nutrients to do the job.

COMMON PESTS

Symptoms	Pest	Description	Treatment
New growth stunted and deformed. Leaves curl, pucker. Stems weak, drooping, spindly.	Aphids	Tiny, soft, light-colored bodies. Sticky honeydew-like substance will be found smeared on leaves.	Remove all damaged growth. Remove affected plants to isolated containers. Rinse foliage in warm water every six days. Spray underside of foliage and stems with Safer's Insecticidal Soap, Pyrethrum, or Malathion.
Gaping holes in leaves. Badly shredded leaves.	Caterpillar	Wormlike body may be hairy, black, brown or some other color (or colors).	Search units thoroughly. Pick off and destroy any caterpillars you find.
Foliage with tips chewed. Gaping holes in foliage.	Cockroaches	Dark brown or black beetlelike bodies.	Remove and destroy any you find on or around plants. Keep the area around your plants free of organic debris. Place roach traps around unit.
Stems, leaves, flowers, fruit with small holes punched in tissue.	Earwigs	Dark brown beetle-like insects about an inch long with a tail appendage resembling a pair of forceps.	Earwigs emerge at night. Use a flashlight to search in and around unit. Use tweezers to pick earwigs off plants and destroy them.

COMMON PESTS

Symptoms	Pest	Description	Treatment
Badly wilted plant. Sudden drop in growth rate. Leaves torn and yellow. Roots become pulpy, as if they've been chewed.	Fungus gnats	Miniscule flying black bodies seen hovering around foliage. Adults harmless. Eggs laid in aggregate hatch into maggots, which attack roots.	Discard badly damaged plants. Flush unit thoroughly with clear water. Spray plants with Pyrethrum or Safer's Insecticidal Soap.
Plants suddenly wilt. Many leaves drop off.	Mealybugs	Tiny insects form white, powdery masses on leaf axils, stem joints. Oval bodies covered by a white, sticky substance.	Scrape off powdery masses. Spray with Safer's soap or Diazinon. (Use Diazinon at five-day intervals.) Use a magnifying glass to inspect both the top and underside of the leaves of each plant.
Leaves turn yellow, are sticky to the touch, and fall off. Plant appears badly wilted and droops.	Scale	Appear as tiny knobs or domes that attach themselves to stems or underside of leaves. Soft bodies, waxy texture when probed with fingernail.	Remove any you see. Thoroughly inspect each plant in the unit and remove any affected plant; isolate it by placing it in a room without plants. Wash the plant in warm, soapy water, gently running your fingers over stems and leaves to be certain that all the protuberances are removed. Safer's soap, Malathion or Diazinon can be used.

COMMON PESTS

Symptoms	Pest	Description	Treatment
Tiny, coarse webs spun across the underside of leaves or spanning leaf axils. Leaves develop gray blotches.	Spider mites (also known as mites, red spider)	Insects too small to be seen with the unaided eye (only 1/50th of an inch long)	Remove affected plants from the unit, isolating them. Remove and destroy any mites you can see. Wash each leaf gently with warm water. Spray with Safer's soap, Pyrethrum, or Kelthane.
Pockmarks, pale scars on underside of leaves. Leaves develop blisters. Buds drop off.	Thrips	Small, slender, hard to spot. Leaf foliage speckled with dark blobs of excrement.	Wash the plants with lukewarm water. Remove any of the insects you spot. Spray with Safer's soap, Pyrethrum, or Malathion.
Leaves suddenly yellow, wilt, fall off. Sticky substance on leaves.	Whiteflies	Tiny, white airborne pests. Usually observed hovering in masses around foliage.	These are stubborn pests. Spray with Pyrethrum, Malathion, or Diazinon. Use Pyrethrum every two days, or Malathion or Diazinon every five days. Spray *all* the plants in the unit affected, not just the infested plant.

PLANT DISEASES

Symptoms	Disease	Treatment
Crown and stem tissue turns soft, pulpy, and rots away.	A fungus known as crown and stem rot.	Use a sterile knife to pare away all diseased tissue. Dust incisions with a fungicide. Check to make certain that the aggregate is not becoming water-logged and that the solution is not collecting in stagnant pools.
Leaves coated with downy gray substance. Leaves then curl up, wither, fall off.	Mildew. Occurs when relative humidity is consistently too high and when ventilation is insufficient.	Remove affected leaves. Increase ventilation in area.
Sooty black or grayish white growth on leaves.	Molds. Black mold occurs as a by-product of an aphid, scale or mealybug attack. Gray mold is caused by waterlogged aggregate.	Wash leaves with lukewarm, soapy water if the mold is black, and identify pest population present. Scrape away gray mold. Examine watering practices.
Depressed spots on leaf surface. Leaf tips shrivel, turn brown. Leaves marked by dark bars.	Anthracnose, a fungus infection commonly caused by too much water. Occurring usually only in plants grown in soil.	Remove damaged leaves. Spray foliage of infected plants with a fungicide. Check aggregate to be certain that it is not generally waterlogged.
Grayish white, fuzzy growth occurring on leaves.	Botrytis, a fungus occurring in situations of inadequate ventilation.	Remove affected leaves. Increase ventilation around plants. Uncommon in hydroponic units.
Leaves turn yellow, veins remain green. Leaf ends whiten or turn brown.	Chlorosis, a non-contagious deficiency disease, caused by an excessively alkaline aggregate or an unbalanced nutrient solution.	Chech the pH of the solution and remedy the imbalance. See the section on pH for specifics.
Leaves wilt, turn brown, fall off. Roots are pulpy and discolored.	Root rot, a fungus.	Cut away diseased portions of roots, using a sterile knife. Treat new root ends by dusting them with a fungicide. Discard aggregate.

CONCERNING CLEANLINESS

Keeping your unit and the area around it clean is not just preferable, it is necessary. Remove any fallen leaves or other debris immediately. Before adding any equipment to a unit, wash the new parts thoroughly with warm, soapy water. If you bring in gardening tools you've used elsewhere, disinfect them by rinsing them in rubbing alcohol. (Many gardeners prefer to set aside one set of tools to use only with their hydroponic units.) Always wash your hands before handling plants. Inspect the plants *every day* for signs of trouble. At least once a year, disassemble each hydroponic unit and wash every piece.

Prevention is the best medicine. Debris both attracts and shelters insects. Even a small corner of the unit, if allowed to become dirty, can affect the entire unit. Dirt in and near a unit can also serve as a carrier of mold and mildew spores, fungi, and diseases.

CHILDREN AND PLANTS

Small children are often as resourceful as cats, and they are twice as curious. It's important to keep them well away from your plants unless you are present to supervise—to protect them as much as to protect your unit. A pot or an entire unit could be tipped over by an inopportune touch. More important, some children have a tendency to sample intriguing substances. While this could be only mildly distasteful, at worst the leaf could be poisonous. Dieffenbachia, oleander, and poinsettias are among the plants having poisonous leaves.

Aside from these considerations, children should be encouraged to become acquainted with the principles and practices of gardening. Hydroponic gardening is sufficiently unusual to attract many older children's attention. If nothing else, a small hydroponic unit—even a single plant grown in a flowerpot—makes an excellent science fair project.

The more removed we become from the experience of producing our own crops, the more unrealistic we tend to become both about the work involved and about the availability of food. We expect to see food in our supermarkets, but how it got there, at what cost to the environment, even under what conditions it was produced, are all unknown to us. The conditions affecting agriculture are not likely to improve until sufficient numbers of people want them to improve. By giving our children some understanding of the principles of gardening—and of the wisest, least-abusive methods which can be used—we make them better prepared to face the problems associated with nature and agriculture.

PETS AND PLANTS

Some dogs and many cats are irresistibly attracted to plants being grown indoors. Cats have been known to curl up on the soil surface of a plant being displayed in a sunny window or to use the surface as a litter box. All of these actions are predictably injurious and often fatal to plants. Large dogs, out of an excess of enthusiasm, have been known to inadvertently topple plants over. Even the swipe of a particularly large tail is enough to send some plants onto the floor.

I doubt that changing the medium in which your plants are grown will much affect the degree of interest shown your plants by your pets. However, unless you are growing individual plants in pots filled with aggregate, hydroponic units will be larger and heavier, less likely to topple over. The damp surface may also discourage interest. The only certain way to resolve the problem is to take every step possible to keep the plants out of the reach of your pets. If a cat is particularly persistent, perhaps you can divert its attention by giving it a plant of its own. If the problem is really severe, you may have to move your unit or pots to a spot where your pet cannot get at them.

ON THE USE OF PESTICIDES

We continue to hear disturbing revelations about the harmful effects suspected or known to be caused by various pesticides. These findings and suspicions have helped to create a degree of uncertainty among gardeners and to stir up many rancorous debates between those who favor the use of chemicals and those who do not. One positive outcome of this debate is the caution many gardeners are beginning to display toward the use of pesticides and fungicides.

For many years, the dominant philosophy was to use chemicals liberally whenever there was even a suspicion of trouble. However, now we know that even when we must resort to chemicals, we need to apply only a limited amount. Doubling the amount of chemical applied won't double protection from trouble, or end an insect infestation twice as fast. The more we use, the more we contribute to an environment already overtaxed with chemicals and pollutants.

I use a pesticide or fungicide only when I am convinced there is no alternative. I start by using the mildest chemical remedies available. If any single plant is badly infested with a pest or disease, I get rid of it. It's simpler to replace one plant than to repeatedly treat many.

When applying any insecticide to plants in a hydroponic unit, I recommend using a spray. It's the fastest way to reach the bugs you'll be fighting (insects attacking the roots of plants are uncommon in hydroponic units). Ventilate the area thoroughly when you spray. If you apply any chemical spray to vegetable plants, wait at least two weeks after the last application before harvesting the vegetables. Wash each vegetable *thoroughly* in warm water.

Always use pesticides *only* in well-ventilated areas. It's worth noting that insects hate well-ventilated areas. Areas exposed to a free flow of fresh air are also far less likely to develop problems with a fungus or mold. Keep your indoor gardening areas well-ventilated with a fan and, in temperate weather, open the windows.

Safer's Insecticidal Soap is an excellent organic treatment. Made from fatty acid salts and alcohol, it kills bugs but does not contain any of the poisonous compounds found in chemical pesticides. It must be used frequently and sprayed very thoroughly over all of the plants in a unit.

Pyrethrum, available in an aerosol, is derived from a flower and is a powerful natural insecticide. I have used it successfully to swiftly end a spider mite infestation. Because it decomposes rapidly, you must apply Pyrethrum several times to eliminate a problem.

Kelthane, used frequently by large-scale greenhouse growers, is a *very* potent, very effective poison. It will end an infestation, but must be used with great caution.

Malathion was once the pesticide of choice for gardeners. It is effective against a wide variety of insects—but it is also toxic to humans.

If you have a serious infestation and must use either Kelthane or Malathion, do so with caution. Follow all the recommended procedures. Use only as much as the instructions call for.

FUNGICIDES

Fungicides are chemical preparations designed to destroy disease organisms. They also serve to prevent the reoccurrence of an infection. They are available as soluble powders or as liquids in a concentrated form. Fungicides can be sprayed on foliage or applied to the growing medium. Trade names of fungicides include Ferbam, Maneb, and Zineb. Instructions printed on the containers indicate the diseases against which they are effective.

When preparing insecticides and fungicides:
1. Read carefully the instructions printed on the container and *follow them.*
2. Avoid inhaling the mix when preparing a poison. Don't prepare the mix in an unventilated room. If any of the mix splashes onto your hands or clothes, wash it off immediately.
3. Do not apply a mix while children or animals are close at hand. After the solution has been applied, try to place the plants that have been treated in a spot where they cannot be reached.

4. If you spray a plant, spray it only in a well-ventilated area.

5. Mix only as much of a preparation as you need. Discard any that is left over.

6. Store the poison containers in a cupboard or some area where they cannot be reached by anyone but an adult. Keep the lids of the containers tightly sealed.

7. Use only as much of a preparation as you are instructed to use. *Don't get carried away.*

PESTS AND DISEASES IN THE GREENHOUSE

I can think of few aspects of gardening as frustrating as trying to combat an infestation of pests or an infection damaging the plants being grown in a greenhouse. If you're raising plants in your rooms, any illness or insect attack will be somewhat contained by the distance between plants being grown in pots or other widely spaced containers. If you're growing plants outdoors, the profusion of green life, the actions of the elements, and the presence of insect predators will all help to limit the amount of damage done to any one crop. But in a greenhouse, great numbers of plants are grown close together. There are no barriers to a pest or disease spreading rapidly from plant to plant, and there are no natural predators present to control the levels of pests. In addition, the beneficial growing conditions that a greenhouse provides for plants also serve to encourage the rapid proliferation of some pests and diseases.

The best way to deal with such problems in a greenhouse is to do everything you can to prevent them. Above all, keep your greenhouse clean. Remove any organic debris as soon as you see it. Each time you are in the greenhouse take a close, careful look around for any signs of damage to the plants. You can get a head start on the problem by thoroughly cleaning your greenhouse before a single plant is moved in.

After the greenhouse has been fully assembled, scrub the interior down using any of the common household disinfectant cleansers. Clean the trays, benches or other plant containers before setting them in place, using hot, soapy water to scrub them out and cool clean water to rinse off the suds. Some greenhouse gardeners sterilize all of their equipment with one of the several commercial preparations containing methyl bromide. However, in my opinion such preparations are too dangerous for the novice to use. While I am all for preventing any problems, I don't think it is a wise idea for someone unfamiliar with the handling of powerful poisons to use preparations that, if applied improperly, could have a profoundly nasty effect on oneself, one's family, one's pets, or any otherwise harmless wildlife innocently coming into contact with your greenhouse.

Since you're starting out with a clean greenhouse, you'll already be one step ahead of most problems. You can maintain that lead by keeping the greenhouse clean—and that means every part of it. Remove any webs of dust you discover as soon as you find them. Keep the floor free of any debris. Some serious greenhouse gardeners spray the interior of their units several times a year, using a solution composed of soap and water or of a biodegradable detergent dissolved in water. You could use a sheet of plastic to cover the plants while you're spraying.

If a pest should settle in, don't dawdle about handling the problem. Apply the required pesticide immediately. Discard any badly infested plants. Inspect all the other plants every day (if possible several times a day) to be certain that the problem has not spread. The same procedures hold true for an attack of fungus or mold. Use a fungicide immediately—speed really is of the essence.

Before carrying any tools into the greenhouse, wash them off. And if you've been working with plants in your garden, wash your hands before touching any of the plants in your greenhouse. It takes effort and thought to keep a greenhouse free of pests or diseases. However, it is certainly not impossible to do so. Some greenhouse gardeners go for years without encountering a serious problem. So can you, if you work at it.

WILL HYDROPONICS REDUCE PEST AND DISEASE PROBLEMS?

Yes—in some cases. Plant pests living in soil and soil-borne fungus infections, mildews, and molds are immediately excluded when you switch from gardening in soil to gardening in sterile growing medium. Pests and diseases not normally found in soil may still occur, although they occur less often and when they do occur cause less trouble. It is certainly true that in a properly run hydroponic unit the plants will be very healthy. And healthy plants are more resistant to disease, more capable of recovering from damage caused by insects or infections.

Afterword

It seems inevitable that hydroponics will come to play an increasingly important part in agricultural methods, especially in developing nations and in areas where most fresh foodstuffs must be imported. It seems equally inevitable that hydroponics will become a very popular, very common form of home gardening. And, according to the recent pronouncements of scientists involved in the space program, when humans move into space for long periods of time, they will supply themselves with fresh food by operating hydroponic gardens. Hydroponics is a method whose day is about to come. and that is all to the good.

We need more such innovations—inexpensive enough for even impoverished peoples to implement; simple enough for people without any background in science; and designed wisely enough to be less abusive of the environment, less costly in the use of nonrenewable resources. Hydroponics has all of these qualities, and as such serves as an example of what we need, both in agriculture and in other fields.

There is no end to this book—if it has well served its purpose. It has been my hope to interest you sufficiently in hydroponics to encourage you to pursue the subject, to start experimenting, to continue reading, to work with your own unit. Hydroponics is still in a sufficiently incomplete state to benefit from the work of all who are interested in it. Nonprofessional gardeners have already made significant contributions to the field, and I expect they will continue to do so.

There is one further thing each reader can do. If, after you have experimented with hydroponics, you are convinced of its virtues, don't keep it to yourself. There are many who have not heard of the method or have dismissed it as being of little consequence. No one is more convincing, I think, than a gardener speaking from firsthand experience. So if you believe in hydroponics, let others know.

Pass it on.

Resources

GREENHOUSES AND GREENHOUSE ACCESSORIES

Here are addresses of some of the firms selling greenhouses and greenhouse accessories by mail. Some of these firms, sensitive to trends in gardening, have begun to carry hydroponic units and accessories as well.

These units are prefabricated and shipped unassembled. The degree of skill and the amount of time required to assemble a unit vary widely from model to model. I suggest that you send away for a number of catalogs or price lists. The wide range of models on the market should allow you to find one that will fit your needs and budget.

Bloomin' Greenhouses, Inc.
10909–9 Atlantic Boulevard
Jacksonville, Florida 32225

Aluminum-frame and fiber glass greenhouses, as well as heating and cooling systems and hydroponic units. (Catalog, $3)

Brighton By-Products, Inc.
P.O. Box 23
New Brighton, Pennsylvania 15066

Excellent supplier of horticultural needs with a strong emphasis on greenhouse supplies and nutrients. (Catalog, $5)

Charley's Greenhouse Supplies
1569 Memorial Highway
Mt. Vernon, Washington 98273

Tools, lights, nutrients, greenhouses, and greenhouse accessories. (Catalog, $2.)

Cropking Greenhouses
P.O. Box 310
6142 Wooster Pike
Medina, Ohio 44258

Greenhouses and equipment for the large-scale hydroponic grower. Much of its equipment is adaptable for smaller units. A hydroponics newsletter also is available. (Catalog, $2)

Down-to-Earth Distributors
850 W. 2nd Street
Eugene, Oregon 97402

Organic pest controls and nutrients. (Catalog, free)

Everlite Greenhouses
P.O. Box 11087
Cleveland, Ohio 44111

Aluminum-frame greenhouses and supplies. One of the longtime leaders in greenhouse design. (Catalog, $5)

Gothic Arch Greenhouses
P.O. Box 1564
Mobile, Alabama 36633

Redwood- or cedar-frame greenhouses, in designs reminiscent of the Gothic arch. (Brochure, free)

Gro-Tek
RFD 1, Box 518A
South Berwick, Maine 03908

Wide selection of supplies for indoor and greenhouse gardeners. (Catalog, $1)

Hoop House Greenhouse Kits
Fox Hill Farm
20-N Lawrence Street
Rockville, Connecticut 06066

Simple, inexpensive greenhouse units. (Brochure, $1)

Janco Greenhouses
Dept. NG 589
9390 Davis Avenue
Laurel, Maryland 20707

Large line of greenhouses, solar rooms, and accessories. (Catalog/portfolio, $5)

Santa Barbara Greenhouses 1115-J Avenue Acaso Camarillo, California 93010	Redwood and fiber glass greenhouses, prefabricated and in kits. Also greenhouse accessories. (Catalog, free)
Sturdi-Bilt Mfg. Co. 11304 S.W. Boones Ferry Road Portland, Oregon 97219	Redwood-frame greenhouses and accessories. (Catalog, $2)
Sunglo Solar Greenhouses 4441 26th Avenue West Seattle, Washington 98199	Aluminum-frame greenhouses in a variety of designs and accessories. (Catalog, free)
Turner Greenhouses Route 117 South P.O. Box 1260 Goldsboro, North Carolina 27530	Steel-frame greenhouses in a variety of designs and accessories. (Catalog, free)
Victory Garden Supply 1428 E. High Street Charlottesville, Virginia 22901	Aluminum-frame greenhouses and accessories. (Brochure, free)

HYDROPONIC SUPPLIES

As hydroponic gardening becomes more and more popular, companies are emerging to supply gardeners with both hydroponic units and a wide variety of ingenious accessories. The Hydroponic Society of America (see p. 114) issues a superb list of suppliers to its members. Advertisements for new products continually appear in gardening magazines.

I've listed here firms that I have found to be reliable and responsive. Write as widely as possible for brochures and catalogs. Many such publications are filled with useful information and highly original tips. By comparison shopping you can find what you need at the right price.

Please note: The term "hydroponic system" refers to an entire hydroponic set-up, including some sort of container, growing medium, an irrigation system, and in some cases a lighting system, which firms offer for one price. The term "hydroponic units" refers to hydroponic components sold separately.

Applied Hydroponics
Eastern address:
  208 Route 13
  Bristol, Pennsylvania 19007
Western address:
  3135 Kerner Blvd.
  San Rafael, California 94901

Hydroponic structures and systems, nutrient preparations, lights, and a variety of accessories for hydroponic gardening. (Catalog, free)

Applied Hydroponics of Canada
2215 Walkley
Montreal, PQ, H4B 2J9
Canada

Hydroponic units and supplies, including halide lights. (Catalog in French and English, free)

Aqua Culture
P.O. Box 26467-FG
Tempe, Arizona 85282

Manufacturers of a complete hydroponic system for indoor gardening. (Brochure, free)

Cervantes Indoor Garden Store
9915 S.E. Foster Road
Portland, Oregon 97266

Hydroponic systems, halide and sodium lights, and hydroponic gardening accessories. (Catalog, free)

Floralight Gardens Canada, Inc.
P.O. Box 247, STA. A
Willowdale, Ontario M2N 5S9
Canada

Manufactures Floralight plant stands, useful for indoor propagation of house plants, but adaptable for use with hydroponic units. (Catalog, free)

Full Circle Garden Products
P.O. Box 6
Redway, California 95560

Offers a wide range of nutrients and tools, plant lights, and rockwool, which is the growing medium of choice for commercial hydroponic firms. (Catalog, $2)

Green Thumb Hygro-Gardens
P.O. Box 1314
Sheboygan, Wisconsin 53081

Hydroponic systems and accessories. (Catalog, $1)

Harvest Glow Systems
32 E. Fillmore Avenue
St. Paul, Minnesota 55107

Hydroponic units and hydroponic systems, tools, lights, nutrients, and other accessories. (Catalog, free)

Hollisters Hydroponics
P.O. Box 16601B
Irvine, California 92713

Hydroponic units and supplies.
(Catalog, $1)

Hydro-Gardens of Denver
P.O. Box 9707
Colorado Springs, Colorado
80932

Supplies for the greenhouse as
well as hydroponic units and accessories. (Catalog, free)

LIGHTING SUPPLIES

In addition to the suppliers listed in the Hydroponics Supplies section, the following firms offer a variety of lighting units and accessories for in-door gardening.

Agrilite
P.O. Box 12
93853 River Road
Junction City, Oregon 97448

Halide, sodium, and full-spectrum
discharge lights, as well as other
gardening tools and supplies.
(Catalog, $2)

Geo-Technology
1035 17th Avenue, Dept. H
Santa Cruz, California 95062

High-intensity halide lights. (Catalog, free)

The Green House
1432 W. Kerrick Street
Lancaster, California 93534

Offers the Gro-Cart, a lighted
plant stand for indoor gardening.
(Catalog, free)

Indoor Gardening Supplies
Box 40567H
Detroit, Michigan 48240

Plant stands, light fixtures, lamps,
and accessories, as well as a wide
range of accessories for indoor
gardeners. (Catalog, free)

Public Service Lamp Corporation
410 West 16th Street
New York, New York 10011

Offers mercury vapor lamps
adaptable for use with indoor
gardens. (Catalog, free)

Wendelighting
2445 N. Naomi Street
Burbank, California 91504

Indoor and outdoor lighting systems for gardens. (Catalog, free)

SEEDS AND PLANTS

Every issue of every gardening magazine carries ads from firms selling mail-order seeds and plants. One of the special pleasures of gardening is, I think, the chance to browse through each season's crop of seed and plant catalogs.

Both *Gardening by Mail 2* and *The Gardener's Book of Sources* (see p. 121) contain extensive, descriptive listings of companies selling seeds and plants by mail.

In addition, the National Gardening Association publishes the *Directory of Seed and Nursery Catalogs,* which lists some 400 mail-order sources of vegetable seeds, fruit trees, herbs, and wildflowers. It is available for $4.00 from:

Catalogs
National Gardening Association
180 Flynn Avenue
Burlington, Vermont 05401

SOCIETIES AND ORGANIZATIONS

Because gardeners are such enthusiastic transmitters of information and ideas, there are an extraordinary number of formal and informal organizations devoted to some aspect of gardening.

The Hydroponic Society of America is a great source of information (see p. 114). *The Gardener's Book of Sources* and *Gardening by Mail 2* (see p. 121) carry extensive lists of organizations and information networks. In addition, *North American Horticulture,* published by Scribner's, and available in many libraries, gathers together essential information on national and regional gardening associations and societies, garden clubs, institutions active in gardening research, conservation organizations, state and federal agricultural organizations, horticultural libraries, public gardens and arboretums, museums, community garden programs, and garden and flower shows.

Bibliography

If you find hydroponics intriguing and you'd like to learn more about it, I suggest you continue to read, join a gardening organization, or write to manufacturers for information on their products.

Gardening is perhaps the least secret of avocations, and gardeners are among the most generous of colleagues. There is no lengthy apprenticeship, no adept from whom to solicit secret formulas. You learn to garden best by doing it, by reading, and by listening to fellow gardeners. This section should help you draw upon the vast amount of useful (and constantly updated) information on gardening now available.

I also recommend that you consider joining this organization:

Hydroponic Society of America
P.O. Box 6067
Concord, California 94524

The society publishes a newsletter and offers an excellent list of suppliers of hydroponic equipment and books on the subject. There is an annual conference, and the record of the conference proceedings is an excellent source of new work in the field. For information about membership, write to the society, enclosing a stamped, self-addressed envelope.

In my opinion, reading is an essential gardening activity. We can turn to books when we need an answer to a question or when we simply want to expand our knowledge of the field. A small reference library is vital to any gardener. I urge you to go to the library and start browsing through the gardening section. If your library doesn't have a title you've heard of, they may be able to secure it for you through an inter-library loan.

Below you'll find a list of works in hydroponics and the other basic elements of gardening. You'll find that every good book will steer you to other remarkable books. Some of the books listed in this section may be out of print. However, they are here because I believe them to be some of the best of their kind. Check your library for books that you can't find in a bookstore. In addition to libraries and bookstores, try the several dealers of second-hand gardening books (listed in Gardening by Mail) for any book you'd like to add to your collection.

Remember, reading isn't an alternative to gardening: it's a necessary supplement to gardening. It's the way gardeners communicate what they've learned. When you garden, you become a part of a community that depends on the generous circulation of theories and techniques.

HYDROPONICS

Because the field of hydroponics continues to evolve, most of the books written thus far are either reports of the latest developments or attempts to provide a theoretical framework for the field. A definitive, encyclopedic work on hydroponics has not yet appeared, and it may be several more years before agreement on enough procedures has been reached to allow such a massive work to be completed. Until then, a number of excellent introductory works on hydroponics as well as several theoretical works of interest are available.

The books listed below generally describe hydroponic units more complex than the ones introduced in this book. They explore in greater depth such matters as growing techniques and soilless mediums.

Bridwell, Raymond. *Hydroponic Gardening.* Santa Barbara: Woodbridge Press, 1974.
   This remains an excellent introductory work, written in a casual style that sets one at ease while communicating clearly the essential techniques. The second half of the book is devoted to the specific challenges of hydroponic gardening in greenhouses.

Dickerman, Alexandra. *Discovering Hydroponic Gardening.* Santa Barbara: Woodbridge Press, 1975.

An introductory text notable for its focus on small-scale hydroponics, and for the personal, engaging tone of the text.

Douglas, James Sholto. *Advanced Guide to Hydroponics.* New York: Drake Publishers, 1976.

Douglas is one of the pioneers in the field and one of the few truly international figures associated with hydroponics. Active both as a theoretician and a practitioner, and constantly experimenting and adapting, Douglas continues to contribute to the ever-growing field of hydroponics. This work is an attempt to create a synthesis of knowledge based on research in labs and in the field. This is not a how-to text, but a fascinating theoretical work.

Douglas, James Sholto. *Beginner's Guide to Hydroponics.* New York: Drake Publishers, 1973.

A clear, authoritative introduction, containing more technical background than any other introductory text.

Douglas, James Sholto. *Hydroponics, The Bengal Method.* New York and London: Oxford University Press, 1975.

A classic. Important background reading, this book clearly demonstrates that hydroponics *can* be done without the intensive application of money and scarce resources. Douglas shows how hydroponics can be adapted to the needs of developing nations, and can function in a variety of environmental conditions without elaborate, delicate machinery. This is a book that forcefully reminds us of one of the essential elements of horticultural research: the quest to find ways to help the world feed itself without squandering water and without expensive machinery. Douglas's work in Bengal gave the first irrefutable proof of the usefulness of hydroponics in underdeveloped nations. This book is the record of the trials and ultimate successes of the hydroponic systems he helped to invent.

Harris, Dudley. *Hydroponics: Growing Without Soil.* North Pomfret, Vt.: David and Charles, 1975.

Harris offers details on a wide variety of hydroponic units, with clear, detailed information on both construction and maintenance.

Jones, Lem. *Home Hydroponics...And How to Do It!* Pasadena: Ward Ritchie Press, 1975.

Jones places special emphasis on the construction, outfitting, and management of hydroponic greenhouses. If you want to run a hydroponic greenhouse, read this book.

Mittleider, J.R. *More Food from Your Garden*. Santa Barbara: Woodbridge Press, 1975.

Mittleider explains his method in a text intended for the beginner, with a winning, clear style. *Food for Everyone,* by Mittleider and Andrew Nelson, published by the same press, is the magnum opus of the Mittleider method, containing 600 pages, 1000 drawings and photographs. It is intended for those who want to apply this (controversial) method on a large scale.

Resh, Howard M. *Hydroponic Food Production*. Santa Barbara: Woodbridge Press, 1983.

In my opinion, the most thorough work on hydroponics presently available.

Schubert, M., and Blaicher, W. *The ABC's of Hydroponics*. New York: Sterling Publishers, 1984.

Another clear, thorough text for the beginner.

Sutherland, Struan. *Hydroponics for Everyone*. Melbourne, Australia: Hyland House, 1986.

There is a great deal of enthusiasm for hydroponics in Australia. This work is one of my favorites because it mingles the author's contagious enthusiasm with a great deal of detailed, specific information. This is a perfect model of the genre of gardening books which records a gardener's discovery of a field, including his mistakes and victories.

GARDENING UNDER LIGHTS

Brooklyn Botanic Garden Handbooks. *Gardening under Lights*. New York: Brooklyn Botanic Gardens.

Handbooks issued by the Brooklyn Botanic Gardens are clear, well-illustrated, and specific. This handbook (#93 in the series) covers various light systems, basic techniques, the specific needs of a wide variety of plants, and common problems.

Cervantes, George. *Gardening Indoors: How to Grow with High Intensity Discharge Lamps*. Portland: Interport USA, Inc., 1986.

The best introduction to the use of HID lamps in gardening. Very thorough, clear, and convincing.

Ebert, George A. *The Indoor Light Gardening Book.* New York: Crown Publishers, 1973.

It's too early to say how accessible HID lamps will become. They could revolutionize indoor gardening. In the meantime, Ebert's book remains a useful survey of other methods using artificial light.

Fitch, Charles Marsden. *The Complete Book of Houseplants Under Lights.* New York: Hawthorne Books, 1975.

An excellent explanation of the needs of houseplants grown under lights, with very specific information.

GREENHOUSES

McDonald, Elvin. *How To Build Your Own Greenhouse.* New York: Popular Library, 1976.

How to plan, erect, and maintain a greenhouse—and what to put in it.

Neal, Charles D. *Build Your Own Greenhouse.* Radnor, Pa.: Chilton Book Company, 1975.

Complete, well illustrated, but with an emphasis on large, elaborate models.

Walls, Ian G. *The Complete Book of Greenhouse Gardening.* New York: Quadrangle Books, 1975.

Very thorough coverage of the design and maintenance of greenhouses and the selection and care of crops. An excellent guide and reference.

Yanda, Bill, and Fisher, Rick. *Solar Greenhouse Design, Construction, Operation.* Santa Fe: John Muir Publications, 1976.

A superb work. Concise yet thorough, encouraging, practical, and detailed. The idea of running a greenhouse without high electricity bills is very appealing.

HERBS

Gilbertie, Sal, and Sheehan, Larry. *Herb Gardening at Its Best.* New York: Atheneum/SMI, 1982.

Excellent, thorough introduction.

Kowalchik, C., and Hylton, William, eds. *Rodale's Illustrated Encyclopedia of Herbs.* Emmaus, Pa.: Rodale Press, 1987.

The standard reference: history, uses, and specific cultivation practices; superbly illustrated.

Prenis, John. *Herb Grower's Guide.* Philadelphia: Running Press, 1974.
A clear introduction to the history, cultivation, and uses of twenty-five of the most popular herbs.

Sinclair, Eleanor. *A Garden of Herbs.* New York: Dover Publications, 1969.
A delightful book on the history and immensely varied uses of herbs.

HOUSE PLANTS

Baylis, Maggie. *House Plants for the Purple Thumb.* San Francisco: 101 Productions, 1981.
A witty, clear introduction to house plants for the beginner.

Reader's Digest Success with Houseplants. New York: Random House, 1979.
Detailed information on flowering and foliage plants that can be grown indoors, with a strong emphasis on how to provide adequate environments.

Seddon, George. *Essential Guide to Perfect House Plants.* New York: Summit Books, 1985.
Very useful information on caring for plants indoors.

Wright, Michael. *Complete Indoor Gardener.* New York: Random House, 1975.
This book contains thorough, excellent information on the specific needs of individual species of plants.

PLANT PROBLEMS

Carr, Anne. *Rodale's Color Handbook of Garden Insects.* Emmaus, Pa.: Rodale Press, 1983.
Some 300 varieties of garden pests are profiled here, with photographs of both larval and adult forms, and detailed information on the habits of each species. A variety of organic methods of dealing with each species are given.

Cornell Bulletin 74. *Safe Pest Management around the Home.* Ithaca: New York State College of Agriculture and Life Sciences, 1989.
An excellent guide to the careful use of pesticides inside and around your home. Distinguished by both its clarity and its detail, it contains lengthy lists of plants and the bugs most likely to be attracted to them, as well as specific explanations of how to select and apply pesticides.

The Ortho Problem Solver. Stone Mountain, Ga.: Ortho Information Services, 1989.

Check your local or county library for this one. A massive (1,000+-page, $225) guide to the weeds, diseases, and pests that can wreak havoc in your garden. Superbly illustrated and thorough, but with, not surprisingly, an emphasis on the use of Ortho products in remedying garden problems. Still, a very useful encyclopedia.

Westcott, Cynthia. *The Gardener's Bug Book* and *Westcott's Plant Disease Handbook.* Revised by R. Kenneth Holst. New York: Doubleday, 1979.

These encyclopedic volumes list and thoroughly describe all of the plant pests and diseases commonly found in North America. You may not need to own them, but you are likely at some point to want to refer to them. No matter how exotic the problem, it most often can be found here, described at length, with a variety of treatments clearly given.

PLANT PROPAGATION

Browse, Philip McMillan. *Plant Propagation.* New York: Simon & Schuster, 1988.

A short but thorough, well-illustrated guide to every kind of plant propagation technique.

Bubel, Nancy. *The New Seed Starter's Handbook.* Emmaus, Pa.: Rodale Press, 1989.

Thorough, specific, and clear. Like Browse's *Plant Propagation,* an excellent guide.

Hartman, Hudson T., and Kester, Dale E. *Plant Propagation: Principles and Practices.* Englewood Cliffs, N.J.: Prentice-Hall, 1975.

Either Bubel or Browse should tell you what you need to know to propagate plants effectively.

This book is a more technical, encyclopedic approach to the subject. However, for anything you can't find elsewhere, check here. If it has to do with plant propagation, it will be covered in this massive text.

Nehrling, Arno and Irene. *Propagating House Plants.* New York: Bantam Books, 1976.

A step-by-step guide to this specific subject. It covers everything you need to know to propagate house plants.

SOURCE BOOKS

Barton, Barbara J. *Gardening by Mail 2*. San Francisco: Tusker Press, 1987.

Logan, William Bryant. *The Gardener's Book of Sources*. New York: Penguin, 1988.

Serious gardeners should own these two books. They are both superb guides to the abundance of products and resources available to gardeners, listing (and describing) everything from seed suppliers to manufacturers of gardening furniture, from a supplier of miniature roses to an importer of British gardening tools. The annotations are clear, often witty, and always informative.

Both books allow you to indulge in one of gardening's greatest pleasures: browsing. The catalogs of gardening merchants are usually filled with surprises and delightful ideas. These books are catalogs of catalogs, extraordinary (and informed) guides to the immense variety of resources available to gardeners. I suggest you buy these books. You'll use them again and again. Further testimony to the great variety of products now available to gardeners, there is not a great deal of overlapping in these books. Both books are refreshingly personal in tone and taste.

VEGETABLES

National Gardening Association. *Gardening: The Complete Guide to Growing America's Favorite Fruits and Vegetables*. New York: Addison-Wesley, 1986.

A thorough and clear introduction to food gardening. It not only tells you how to carry out a procedure, it explains why that procedure is necessary and how it is effective. If you want just one book on the subject, I'd recommend this title.

Raymond, Dick. *Joy of Gardening*. Pownal, Vt.: Storey Communications, 1983.

A personable, complete guide to the subject, distinguished by the author's warm tone and often ingenious solutions to problems in the garden.

Index